Climate 2030

A NATIONAL BLUEPRINT FOR A CLEAN ENERGY ECONOMY

Rachel Cleetus

Steven Clemmer

David Friedman

Union of Concerned Scientists

Citizens and Scientists for Environmental Solutions

MAY 2009

Rachel Cleetus is an economist with the Union of Concerned Scientists
Climate Program.
Steven Clemmer is the research director of the Union of Concerned
Scientists Clean Energy Program.
David Friedman is the research director of the Union of Concerned
Scientists Clean Vehicles Program.

The Union of Concerned Scientists (UCS) is the leading science-based
nonprofit working for a healthy environment and a safer world.

More information about the Union of Concerned Scientists is available
on the UCS website at *www.ucsusa.org*.

The full text of this report and additional technical appendices are
available on the UCS website (*www.ucsusa.org/blueprint*) or may be
obtained from:

UCS Publications
2 Brattle Square
Cambridge, MA 02238-9105

Or, email *pubs@ucsusa.org* or call (617) 547-5552.

Designed by:
DG Communications, Acton, MA
www.NonprofitDesign.com

Printed on recycled paper.

Table of Contents

FIGURES, TABLES, AND BOXES

Figures

Tables

Boxes

Acknowledgments

The production of this report was made possible in part through the generous support of Avocet Charitable Lead Unitrust, David and Leigh Bangs, The Educational Foundation of America, The Energy Foundation, Foundation M, Bernard F. and Alva B. Gimbel Foundation, Ben and Ruth Hammett, The William and Flora Hewlett Foundation, The Dirk and Charlene Kabcenell Foundation, Korein Foundation, The John Merck Fund, Oak Foundation, Scherman Foundation, Inc., NoraLee and Jon Sedmark, Wallace Genetic Foundation, Inc., and Wallace Global Fund.

For independent expert review of the report, the authors thank Frank Ackerman (Stockholm Environment Institute–U.S. Center), Jeff Alson (U.S. Environmental Protection Agency), Doug Arent (National Renewable Energy Laboratory), Lynn Billman (National Renewable Energy Laboratory), Peter Bradford (Vermont Law School), John Byrne (Center for Energy and Environmental Policy, University of Delaware), Duncan Callaway (Center for Sustainable Systems, University of Michigan), Elizabeth Doris (National Renewable Energy Laboratory), Paul R. Epstein (Center for Health and the Global Environment, Harvard Medical School), John German (International Council on Clean Transportation), Jeffery Greenblatt (Google.org), Christopher A. James (Synapse Energy Economics), Erin Kassoy (The Alliance for Climate Protection), David Kline (National Renewable Energy Laboratory), Chuck Kutscher (National Renewable Energy Laboratory), Daniel A. Lashof (Natural Resources Defense Council), Brenda Lin (American Association for the Advancement of Science), Thomas R. Mancini (Sandia National Laboratories), Jason Mark (The Energy Foundation), William Moomaw (Fletcher School, Tufts University), Dean M. Murphy, Brian Murray (Nicholas Institute, Duke University), Gregory Nemet (University of Wisconsin–Madison), Joan Ogden (University of California–Davis), Steven E. Plotkin (Argonne National Lab), William H. Schlesinger (Cary Institute of Ecosystem Studies), Monisha Shah (National Renewable Energy Laboratory), Daniel Sperling (University of California–Davis), Laura Vimmerstedt (National Renewable Energy Laboratory), and Michael P. Walsh (international consultant).

We also thank Marilyn A. Brown (Georgia Institute of Technology), Ryan Wiser (Lawrence Berkeley National Laboratory), and several anonymous independent experts for their thoughtful comments during a preliminary presentation of our results.

Organizational affiliations are listed for identification purposes only.
The opinions expressed in this report are solely the responsibility of the authors.

Modeling support was provided by OnLocation, Inc. John "Skip" Laitner and others at the American Council for an Energy-Efficient Economy provided an analysis of the impact of greater energy efficiency in industry and buildings. Marie Walsh at the University of Tennessee provided an analysis of biomass potential. We also thank Nora Greenglass and Richard A. Houghton (Woods Hole Research Center) and Steven Rose (Global Climate Change Research Group, Electric Power Research Institute) for providing expertise in the agriculture and forestry sectors.

This report is the result of years of dedication from a large interdisciplinary team at UCS. The authors thank the combined leadership of Nancy Cole, Peter Frumhoff, Kevin Knobloch, Alden Meyer, Alan Nogee, Lance Pierce, Kathleen Rest, Michelle Robinson, Suzanne Shaw, and Lexi Shultz. Several other UCS experts provided analytic and technical support throughout the process, including but not limited to Don Anair, Doug Boucher, Brenda Ekwurzel, Melanie Fitzpatrick, Kristen Graf, Jeremy Martin, Margaret Mellon, Patricia Monahan, and John Rogers. We wish to give a special thanks to energy analysts Jeff Deyette and Sandra Sattler for their significant analytic support, technical contributions, and constant troubleshooting. Policy expertise, guidance, and additional contributions were provided by Kate Abend, Ron Burke, Christopher Busch, Cliff Chen, Barbara Freese, Eli Hopson, Aaron Huertas, Jim Kliesch, Rouwenna Lamm, Ben Larson, Claudio Martinez, Lena Moffitt, Scott Nathanson, Lisa Nurnberger, Liz Martin Perera, Spencer Quong, Ned Raynolds, Emily Robinson, Erin Rogers, Jean Sideris, Ellen Vancko, Marchant Wentworth, Laura Wisland, and former UCS vehicles engineer Don MacKenzie.

Bryan Wadsworth and Heather Tuttle helped to coordinate the production process, and we thank Sandra Hackman for her tremendous editing work within a tight time frame, and David Gerratt of DG Communications for his design of the report.

Thank you to Julie Ringer for her administrative and production assistance, and to Eric Misbach for his support and assistance to Rachel Cleetus and the UCS leadership in managing this complex project.

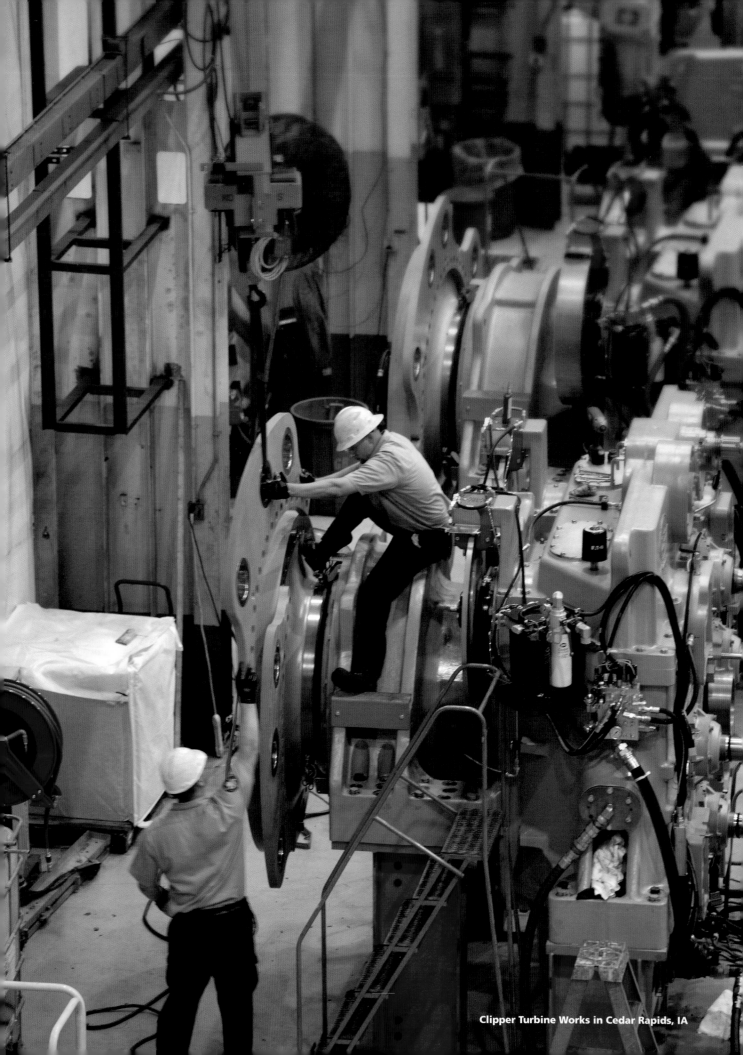

Clipper Turbine Works in Cedar Rapids, IA

Executive Summary

REDUCING OIL DEPENDENCE. Strengthening energy security. Creating jobs. Tackling global warming. Addressing air pollution. Improving our health. The United States has many reasons to make the transition to a clean energy economy. What we need is a comprehensive set of smart policies to jump-start this transition without delay and maximize the benefits to our environment and economy. *Climate 2030: A National Blueprint for a Clean Energy Economy* ("the Blueprint") answers that need.

Recent rapid growth of the wind industry (developers have installed more wind power in the United States in the last two years than in the previous 20) and strong sales growth of hybrid vehicles show that the U.S. transformation to a clean energy economy is already under way. However, these changes are still too

The Climate 2030 Blueprint shows that deep emissions cuts can be achieved while saving U.S. consumers and businesses $464 billion in 2030.

gradual to address our urgent need to reduce heat-trapping emissions to levels that are necessary to protect the well-being of our citizens and the health of our environment.

Global warming stems from the release of carbon dioxide and other heat-trapping gases into the atmosphere, primarily when we burn fossil fuels and clear forests. The problems resulting from the ensuing carbon overload range from extreme heat, droughts, and storms to acidifying oceans and rising sea levels. To help avoid the worst of these effects, the United States must play a lead role and begin to cut its heat-trapping emissions today—and aim for at least an 80 percent drop from 2005 levels by 2050.

The Blueprint Cuts Carbon Emissions and Saves Money

Blueprint policies lower U.S. heat-trapping emissions to meet a cap set at 26 percent below 2005 levels in 2020, and 56 percent below 2005 levels in 2030 (see Figure ES.2). The actual year-by-year emissions reductions differ from the levels set in the cap because firms have the flexibility to over-comply with the cap in early years, bank allowances, and then use them to meet the cap requirements in later years.

To meet the cap, the cumulative *actual* emissions must equal the cumulative tons of emissions set by the cap. In 2030, we achieve this goal.

The nation achieves these deep cuts in carbon emissions while saving consumers and businesses

FIGURE ES.1. The Sources of U.S. Heat-Trapping Emissions in 2005

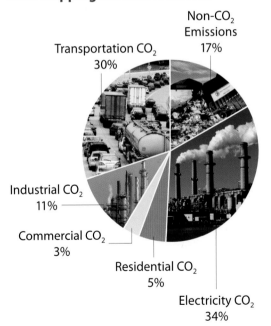

Non-CO$_2$ Emissions 17%

Transportation CO$_2$ 30%

Industrial CO$_2$ 11%

Commercial CO$_2$ 3%

Residential CO$_2$ 5%

Electricity CO$_2$ 34%

The United States was responsible for approximately 7,180 million metric tons CO$_2$ equivalent of heat-trapping emissions in 2005, the baseline year of our analysis. Most of these emissions occur when power plants burn coal or natural gas and vehicles burn gasoline or diesel. The transportation, residential, commercial, and industrial shares represent direct emissions from burning fuel, plus "upstream" emissions from producing fuel at refineries.

Data source: EIA 2008.

The Climate 2030 Approach

This report analyzes the economic and technological feasibility of meeting stringent targets for reducing global warming emissions, with a cap set at 26 percent below 2005 levels by 2020, and 56 percent below 2005 levels by 2030. Meeting this cap means the United States would limit total emissions—the crucial measure for the climate—to 180,000 million metric tons carbon dioxide equivalent (MMTCO$_2$eq) from 2000 to 2030.[a]

The nation's long-term carbon budget for 2000 to 2050—as defined in a previous UCS analysis (Luers et al. 2007)—is 160,000 to 265,000 MMTCO$_2$eq. The 2000–2030 carbon budget in our analysis would put us on track to reach the mid-range of that long-term budget by 2050, if the nation continues to cut emissions steeply.

To reach the 2020 and 2030 cap and carbon budget targets, the Blueprint proposes a comprehensive policy approach (the "Blueprint policies") that combines an economywide cap-and-trade program with complementary policies. This approach finds cost-effective ways to reduce fossil fuel emissions throughout our economy—including in industry, buildings, electricity, and transportation—and to store carbon through agricultural activities and forestry.

Our analysis relies primarily on a modified version of the U.S. Department of Energy's National Energy Modeling System (referred to as UCS-NEMS). We supplemented that model with an analysis of the impact of greater energy efficiency in industry and buildings by the American Council for an Energy-Efficient Economy. We also worked with researchers at the University of Tennessee to analyze the potential for crops and residues to provide biomass energy. We then combined our model with those studies to capture the dynamic interplay between energy use, energy prices, energy investments, and the economy while also considering competition for limited resources and land.

Our analysis explores two main scenarios. The first—which we call the Reference case—assumes no new climate, energy, or transportation policies beyond those already in place as of October 2008.[b] The second—the Blueprint case—examines an economywide cap-and-trade program, plus a suite of complementary policies to boost energy efficiency and the use of renewable energy in key economic sectors: industry, buildings, electricity, and transportation. Our analysis also includes a third "sensitivity" scenario that strips out the policies targeted at those sectors, which we refer to as the No Complementary Policies case.

Our analysis shows that the technologies and policies pursued under the Blueprint produce dramatic changes in energy use and cuts in carbon emissions. The analysis also shows that consumers and businesses reap significant net savings under the comprehensive Blueprint approach, while the nation sees strong economic growth.

a   This amount is equivalent to the emissions from nearly 1 billion of today's U.S. cars and trucks over the same 30-year period. The nation now has some 230 million cars and trucks, and more than 1 billion vehicles are on the road worldwide. Given today's trends, we can expect at least 2 billion vehicles by 2030 (Sperling and Gordon 2009).

b   Our analysis includes the tax credits and incentives for energy technologies included in the October 2008 Economic Stimulus Package (H.R. 6049), as well as the transportation and energy policies in the 2007 Energy Independence and Security Act. However, the timing of the February 2009 American Recovery and Reinvestment Act did not allow us to incorporate its significant additional incentives.

$464 billion annually by 2030. The Blueprint also builds $1.7 trillion in net cumulative savings between 2010 and 2030.[1]

Blueprint policies stimulate significant consumer, business, and government investment in new technologies and measures by 2030. The resulting savings on energy bills from reductions in electricity and fuel use more than offset the costs of these additional investments. The result is net annual savings for households, vehicle owners, businesses, and industries of $255 billion by 2030.[2]

We included an additional $8 billion in govern-

1   Unless otherwise noted, all amounts are in 2006 dollars, and cumulative figures are discounted using a 7 percent real discount rate.

2   Net savings include both energy bills (the direct cost of energy such as diesel, electricity, gasoline, and natural gas) and the cost of purchasing more efficient energy-consuming products such as appliances and vehicles. The cost of carbon allowances passed through to consumers and businesses is also included in their energy bills.

FIGURE ES.2. Net Cuts in Global Warming Emissions under the Climate 2030 Blueprint

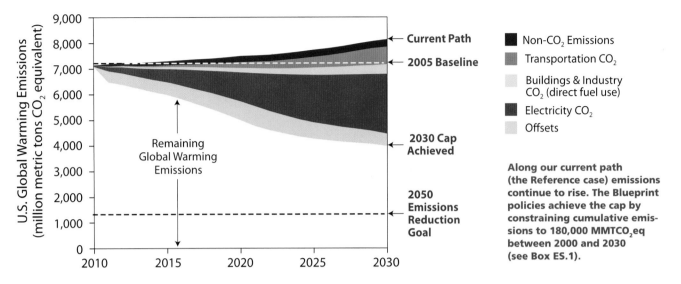

Legend:
- Non-CO$_2$ Emissions
- Transportation CO$_2$
- Buildings & Industry CO$_2$ (direct fuel use)
- Electricity CO$_2$
- Offsets

Along our current path (the Reference case) emissions continue to rise. The Blueprint policies achieve the cap by constraining cumulative emissions to 180,000 MMTCO$_2$eq between 2000 and 2030 (see Box ES.1).

ment-related costs to administer and implement the policies. However, auctioning carbon allowances will generate $219 billion in revenues that is invested back into the economy.[3] This brings annual Blueprint savings up to $464 billion by 2030.[4]

Under the Blueprint, every region of the country stands to save billions (see Figure ES.3). Households and businesses—even in coal-dependent regions—will share in these savings.

The Blueprint keeps carbon prices low. Under the Blueprint, the price of carbon allowances starts at about $18 per ton of CO$_2$ in 2011, and then rises to $34 in 2020, and to $70 in 2030 (all in 2006 dollars). Those prices are well within the range that other analyses find, despite our stricter cap on economywide emissions.

In addition, the Blueprint achieves much larger cuts in carbon emissions *within the capped sectors* because of the tighter limits that we set on "offsets"[5] and because of our more realistic assumptions about the cost-effectiveness of investments in energy efficiency and renewable energy technologies.

The economy grows at least 81 percent by 2030 under the Blueprint. U.S. gross domestic product

Roughly 85,000 people were employed in the wind industry in 2008. In general, renewable energy projects can create more jobs per kilowatt-hour than coal and natural gas power plants.

---

3 We could not model a targeted way of recycling these revenues. The preferred approach would be to target revenues from auctions of carbon allowances toward investments in energy efficiency, renewable energy, and protection for tropical forests, as well as transition assistance to consumers, workers, and businesses in moving to a clean energy economy. However, limitations in the NEMS model prevented us from directing auction revenues to specific uses. Instead, we could only recycle revenues in a general way to consumers and businesses.

4 Values may not sum properly due to rounding.

5 In a cap-and-trade system, rather than cutting their emissions directly, capped companies can "offset" them by paying uncapped third parties to reduce their emissions instead. The cap-and-trade program we modeled includes offsets from storing carbon in domestic soils and vegetation—set at a maximum of 10 percent of the emissions cap, to encourage "decarbonization" of the capped sectors—and from investing in reductions in other countries, mainly from preserving tropical forests, set at a maximum of 5 percent of the emissions cap.

FIGURE ES.3. Net Consumer and Business Savings
(by Census Region in 2030, in 2006 dollars)

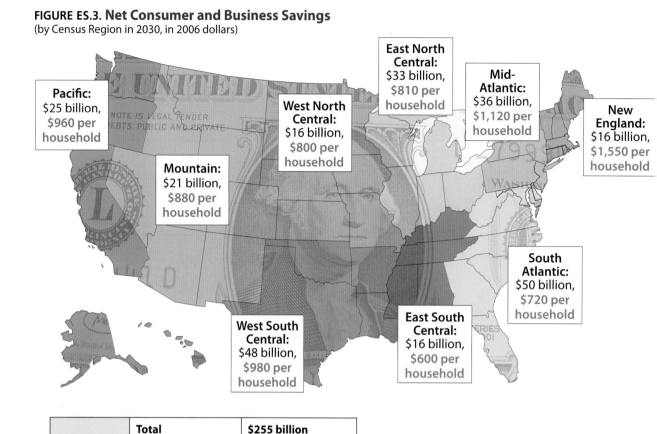

Pacific:
$25 billion,
$960 per
household

Mountain:
$21 billion,
$880 per
household

**West North
Central:**
$16 billion,
$800 per
household

**East North
Central:**
$33 billion,
$810 per
household

**Mid-
Atlantic:**
$36 billion,
$1,120 per
household

**New
England:**
$16 billion,
$1,550 per
household

**South
Atlantic:**
$50 billion,
$720 per
household

**West South
Central:**
$48 billion,
$980 per
household

**East South
Central:**
$16 billion,
$600 per
household

Net Annual Savings in 2030	Total	$255 billion
	Business	$128 billion
	Consumers	$126 billion
	Average Consumer	**$900 per household**

Note: Values may not sum properly because of rounding.

(GDP) expands by 81 percent between 2005 and 2030 under our approach—virtually the same as in the Reference case, which shows the U.S. economy growing by 84 percent. In fact, our model predicts that the Blueprint will slow economic growth by less than 1.5 percent in 2030—equivalent to only 10 months of economic growth over the 25-year period.[6]

The Blueprint also shows practically the same employment trends as the Reference case. In fact, nonfarm employment is slightly higher under the Blueprint than in the Reference case (170 million jobs versus 169.4 million in 2030).

We should note that there are significant limitations in the way NEMS accounts for the GDP and employment effects of the Blueprint policies. NEMS does not fully consider the economic growth that would arise from investments in clean technology, or from the spending

of the money consumers and businesses saved on energy due to these investments. And the Reference case does not include the costs of global warming itself.

The Blueprint cuts the annual household cost of energy and transportation by $900 in 2030. The average U.S. household would see net savings on electricity, natural gas, and oil of $320 per year compared with the Reference case, after paying for investments in new energy efficiency and low-carbon technologies.

Transportation expenses for the average household would fall by about $580 per year in 2030. Those savings take into account the higher costs of cleaner cars and trucks, new fees used to fund more public transit, and declining use of gasoline.

Businesses save nearly $130 billion in energy-related expenses annually by 2030 under the Blueprint. Neither the energy nor the transportation savings account

---

6   This means that under the Blueprint the economy reaches the same level of economic growth in October 2030 as the Reference case reaches in January 2030.

for the revenue from auctioning carbon allowances that will be invested back into the economy, lowering consumer and business costs (or increasing consumer and business savings) even further.

The Blueprint Changes the Energy We Use

Blueprint policies reduce projected U.S. energy use by one-third by 2030. Significant increases in energy efficiency across the economy and reductions in car and truck travel drive down energy demand and carbon emissions.

Carbon-free electricity and low-carbon fuels together make up more than one-third of the remaining U.S. energy use by 2030. A significant portion of

In 2030, the Blueprint cuts the use of petroleum products by 6 million barrels a day— as much as we import from OPEC countries.

U.S. reductions in carbon emissions in 2030 comes from a 25 percent increase in the use of renewable energy from wind, solar, geothermal, and bioenergy under the Blueprint. Carbon emissions are also kept low because the use of nuclear energy and hydropower—which do not directly produce carbon emissions—remain nearly the same as in the Reference case.

The Blueprint reduces U.S. dependence on oil and oil imports. By 2030, the Blueprint cuts the use of oil and other petroleum products by 6 million barrels per day, compared with 2005. That is as much oil as the nation now imports from the 12 members of OPEC (the Organization of Petroleum Exporting Countries). Those reductions will help drop imports to less than 45 percent of the nation's oil needs, and cut projected expenditures on those imports by more than $85 billion in 2030, or more than $160,000 per minute.

Smart Energy and Transportation Policies Are Essential for the Greatest Savings

Many of the Blueprint's complementary policies have a proven track record at state and federal levels. These policies include emission standards for vehicles and fuels, energy efficiency standards for appliances, buildings, and industry, and renewable energy standards for electricity (see Box ES.2). The Blueprint also relies on innovative policies to reduce the number of miles people travel in their cars and trucks.

BOX ES.2.

Climate 2030 Blueprint Policies

Climate Policies

Economywide cap-and-trade program with:

- Auctioning of all carbon allowances
- Recycling of auction revenues to consumers and businesses[*]
- Limits on carbon "offsets" to encourage "decarbonization" of the capped sectors
- Flexibility for capped businesses to over-comply with the cap and bank excess carbon allowances for future use

Industry and Buildings Policies

- An energy efficiency resource standard requiring retail electricity and natural gas providers to meet efficiency targets
- Minimum federal energy efficiency standards for specific appliances and equipment
- Advanced energy codes and technologies for buildings
- Programs that encourage more efficient industrial processes
- Wider reliance on efficient systems that provide both heat and power
- R&D on energy efficiency

Electricity Policies

- A renewable electricity standard for retail electricity providers
- R&D on renewable energy
- Use of advanced coal technology, with a carbon-capture-and-storage demonstration program

Transportation Policies

- Standards that limit carbon emissions from vehicles
- Standards that require the use of low-carbon fuels
- Requirements for deployment of advanced vehicle technology
- Smart-growth policies that encourage mixed-use development, with more public transit
- Smart-growth policies that tie federal highway funding to more efficient transportation systems
- Pay-as-you-drive insurance and other per-mile user fees

---

* See footnote 3.

FIGURE ES.4. Net Cumulative Savings (2010–2030)

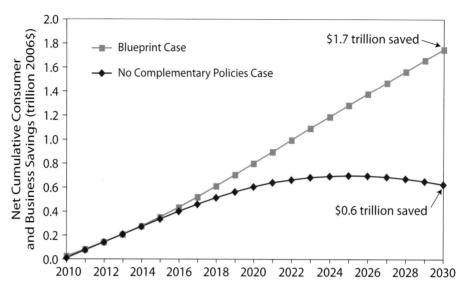

The 2010–2030 net cumulative savings to consumers and businesses are $1.7 trillion under the Blueprint case. Under the No Complementary Policies case, which strips out all the energy and transportation policies, these savings are $0.6 trillion.

*Net present value using a 7% real discount rate

Low-carbon electricity generation is essential if the United States is to meet a stringent cap (or limit) on carbon emissions. Electricity generated from the sun, for example, can power homes and businesses without releasing heat-trapping carbon emissions into the atmosphere. This concentrated solar power plant is one of several renewable energy technologies available to meet the country's electricity needs.

These policies are essential to delivering significant consumer and business savings under the Blueprint. Our No Complementary Policies case shows that if we remove these policies from the Blueprint, consumers

The Blueprint cuts carbon emissions from power plants 84 percent below 2005 levels by 2030.

and businesses will save much less money.[7] Excluding the complementary policies we recommend for the energy and transportation sectors would reduce net cumulative consumer and business savings through 2030 from a total of $1.7 trillion to $0.6 trillion (see Figure ES.4).

Our No Complementary Policies case also shows that excluding the policies we recommend for the energy and transportation sectors will double the price of carbon allowances.

Where the Blueprint Cuts Emissions and Saves Money

Five sectors of the U.S. economy account for the majority of the nation's heat-trapping emissions: electricity, transportation, buildings (commercial and residential), industry, and land use. Blueprint policies ensure that each of these sectors contributes to the drop in the nation's net carbon emissions.

---

7   Some or all of the economic benefits of the complementary policies could also occur if policy makers effectively use the revenues from auctioning carbon allowances to fund the technologies and measures included in these policies. Our study did not address that approach.

FIGURE ES.5. The Source of Cuts in Global Warming Emissions in 2030
(Blueprint case vs. Reference case)

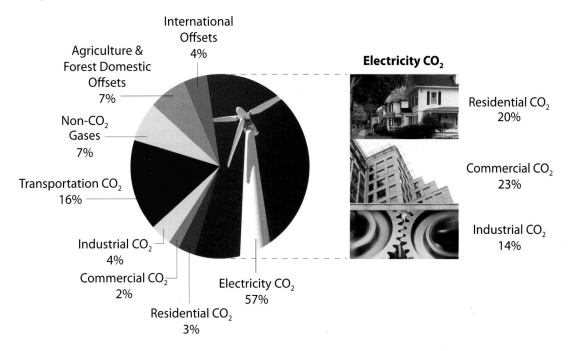

Electricity CO$_2$

Residential CO$_2$
20%

Commercial CO$_2$
23%

Industrial CO$_2$
14%

International Offsets 4%

Agriculture & Forest Domestic Offsets 7%

Non-CO$_2$ Gases 7%

Transportation CO$_2$ 16%

Industrial CO$_2$ 4%

Commercial CO$_2$ 2%

Residential CO$_2$ 3%

Electricity CO$_2$ 57%

TABLE ES.1. Annual Consumer and Business Savings
(in billions of 2006 dollars)

ENERGY SAVINGS	2015	2020	2025	2030
Energy Bill Savings	$ 39	$ 152	$ 271	$ 414
Energy Investment Costs	-40	-80	-123	-160
Net Consumer and Business Savings	$-1B	$72B	$147B	$255B

Energy bill savings include the costs of renewable electricity, carbon capture and storage, and renewable fuels that are passed on to consumers and businesses on their energy bills. Energy investments costs include the cost of more efficient appliances and buildings, cleaner cars and trucks, and a more efficient transportation system.

Note: Values may not sum properly because of rounding.

The electricity sector—with help from efficiency improvements in industry and buildings—leads the way by providing more than half (57 percent) of the needed cuts in heat-trapping emissions by 2030. Transportation delivers the next-largest cut (16 percent). Carbon offsets provide 11 percent of the overall cuts in carbon emissions by 2030. Reduced emissions of heat-trapping gases other than carbon dioxide (non-CO$_2$ emissions) deliver another 7 percent of the cuts. Savings in direct fuel use in the residential, commercial, and industrial sectors are the final pieces, contributing 3 percent, 2 percent, and 4 percent, respectively, of the reductions in emissions (see Figure ES.5).

National savings on annual energy bills (the money consumers save on their monthly electricity bills or gasoline costs, for example) total $414 billion in 2030.

As noted, these savings more than cover the costs of carbon allowances that utilities and fuel providers pass through to households and businesses in higher energy prices. The incremental costs of energy investments (expenditures on energy-consuming products such as homes, appliances, and vehicles) reach $160 billion. The result is net annual savings of $255 billion for households and businesses in 2030.

Households and businesses that rely on the transportation sector see nearly half of the net annual savings ($119 billion) in 2030. However, Blueprint policies ensure that consumers and businesses throughout the economy save money on energy expenses. Lower electricity costs for industrial, commercial, and residential customers are responsible for $118 billion in net annual savings (see Figure ES.6).

FIGURE ES.6. The Source of Savings in 2030
(Blueprint case vs. Reference case)

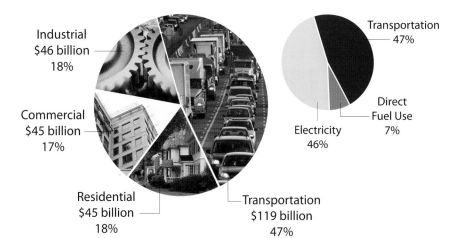

Industrial
$46 billion
18%

Commercial
$45 billion
17%

Residential
$45 billion
18%

Transportation
$119 billion
47%

Transportation
47%

Direct
Fuel Use
7%

Electricity
46%

Consumers and businesses see $255 billion in net annual savings in 2030 under the Blueprint (in 2006 dollars). Consumers and businesses in the transportation sector reap the largest share. Residential, commercial, and industrial consumers each gain just under 20 percent of the net savings, with nearly 90 percent of that amount— or $118 billion—stemming from lower electricity costs.

The MAX light-rail system in downtown Portland, OR, helps residents avoid traffic while commuting to work or going shopping. With smart transportation policies, we can more than double the carbon emissions reductions projected from the 2007 Energy Independence and Security Act (EISA)—ensuring that people from Portland to Pensacola have greater access to mass transit as well as cleaner cars and fuels.

The Blueprint Cuts Emissions in Each Sector

Blueprint policies dramatically reduce carbon emissions from power plants. Under the Blueprint, carbon emissions from power plants are 84 percent below 2005 levels by 2030. Sulfur dioxide (SO_2), nitrogen oxides (NOx), and mercury pollution from power plants are also significantly lower, improving air and water quality and providing important public health benefits.

Most of these cuts in emissions come from reducing the use of coal to produce electricity through greater use of energy efficiency and renewable energy technologies. For example, energy efficiency measures— such as advanced buildings and industrial processes— and high-efficiency appliances, lighting, and motors reduce demand for electricity by 35 percent below the Reference case by 2030. The use of efficient combined-heat-and-power systems that rely on natural gas in the commercial and industrial sectors more than triples over current levels, providing 16 percent of U.S. electricity by 2030. And largely because of a national renewable electricity standard, wind, solar, geothermal, and bioenergy provide 40 percent of the remaining electricity.

Hydropower and nuclear power continue to play important roles, generating slightly more carbon-free electricity in 2030 than they do today. Efforts to capture and store carbon from advanced coal plants, and new advanced nuclear plants, play a minor role, as our analysis shows they will not be economically competitive with investments in energy efficiency and many renewable technologies. However, carbon capture and storage and advanced nuclear power could play a more significant role both before and after 2030 if their costs decline faster than expected, or if the nation does not pursue the vigorous energy efficiency and renewable energy policies and investments we recommend.

Industry and buildings cut fuel use through greater energy efficiency. By 2030, a drop in direct fuel use in industry and buildings accounts for 9 percent of the cuts in carbon emissions from non-electricity sources under the Blueprint.

Transportation gets cleaner, smarter, and more efficient. Under the Blueprint, carbon emissions from cars and light trucks are 40 percent below 2005 levels by 2030. Global warming emissions from freight trucks hold steady despite a more than 80 percent growth in the nation's economy. However, carbon emissions from airplanes continue to grow nearly unchecked, pointing to the need for specific policies targeting that sector. Overall, carbon emissions from the transportation sector fall 19 percent below 2005 levels by 2030—and more than 30 percent below the Reference case.

Much of the improvement in this sector comes from greater vehicle efficiency and the use of the lowest-carbon fuels, such as ethanol made from plant cellulose. Measures to encourage more efficient travel options—such as per-mile insurance and congestion fees, and more emphasis on compact development linked to transit—also provide significant reductions. Renewable electricity use in advanced vehicles such as plug-in hybrids begins to grow significantly by 2030.

These advances represent the second half of an investment in a cleaner transportation system that began with the 2007 Energy Independence and Security Act.[8] These investments provide immediate benefits and will be essential to dramatically cutting carbon emissions from the transportation sector by 2050.

Blueprint Cuts Are Conservative and Practical

The Blueprint includes only technologies that are commercially available today, or that will very likely be available within the next two decades. Our analysis excludes many promising technologies, or assumes they will play only a modest role by 2030. We also did not analyze the full potential for storing more carbon in U.S. agricultural soils and forests, although studies show that such storage could be significant.

Our estimates of cuts in carbon emissions are therefore conservative. More aggressive policies and larger investments in clean technologies could produce even deeper U.S. reductions.

Beyond the Climate 2030 Blueprint—Technologies for Our Future

Our analysis did not include several renewable energy and transportation sector technologies that are at an early stage of development, but offer promise. These include:

- Thin film solar
- Biopower with carbon capture and storage
- Advanced geothermal energy
- Wave and tidal power
- Renewable energy heating and cooling
- Advanced storage and smart grid technologies
- Dramatic expansion of all-electric cars and trucks
- High-speed electric rail
- Expanded public-transit-oriented development
- Breakthroughs in third-generation biofuels

If properly managed, forests and soils are able to store carbon and help reduce the atmospheric carbon overload. Numerous studies suggest significant potential for carbon storage in U.S. agricultural soils and forests. We were unable to analyze the full potential of this option for the Blueprint, suggesting that even greater emissions reductions and savings may be possible.

---

8   Because our Reference case includes the policies in the 2007 legislation, the Blueprint's 30 percent reduction from that case in 2030 represents benefits beyond those delivered from the fuel economy standards and renewable fuel standard in the act. If our Reference case did not include the provisions in the act, Blueprint transportation policies would deliver nearly a 40 percent reduction compared with the Reference case.

The Climate 2030 Blueprint demonstrates that the cost-effective way to achieve quick and deep carbon emissions reductions in the United States is an approach that combines a well-designed cap-and-trade system with policies that promote energy efficiency, the development of renewable electricity, cleaner cars and fuels, and more options to help avoid traffic congestion. The cap ensures the necessary level of emissions reductions while the sector policies help consumers and businesses save more money.

Recommendations: Building Blocks for a Clean Energy Future

Given the significant savings under the Blueprint, building a clean energy economy not only makes sense for our health and well-being and the future of our planet, but is clearly also good for our economy. However, the nation will only realize the benefits of the Climate 2030 Blueprint if we quickly put the critical policies in place—some as soon as 2010. All these policies are achievable, but near-term action is essential.

An important first step is science-based legislation that would enable the nation to cut heat-trapping emissions at least 35 percent below 2005 levels by 2020,[9] and at least 80 percent by 2050. Such legislation would include a well-designed cap-and-trade program that guarantees the needed emission cuts and does not include loopholes, such as "safety valves" that prevent the free functioning of the carbon market.

Equally important, policy makers should require greater energy efficiency and the use of renewable energy in industry, buildings, and electricity. Policy makers should also require and provide incentives for cleaner cars, trucks, and fuels and better alternatives to car and truck travel.

U.S. climate policy must also have an international dimension. That dimension should include funding the preservation of tropical forests, sharing energy efficiency and renewable energy technologies with developing nations, and helping those nations adapt to the unavoidable effects of climate change.

Conclusion

We are at a crossroads. The Reference case shows that we are on a path of rising energy use and heat-trapping emissions. We are already seeing significant impacts from this carbon overload, such as rising temperatures and sea levels and extreme weather events. If such emissions continue to climb at their current rate, we could reach climate "tipping points" and face irreversible changes to our planet.

In 2007 the Intergovernmental Panel on Climate Change (IPCC) found it "unequivocal" that Earth's climate is warming, and that human activities are the primary cause (IPCC 2007). The IPCC report

---

9   Note that this recommendation encompasses more possibilities for reducing emissions than we were able to model in UCS-NEMS. For example, investments in reducing emissions from tropical deforestation could help meet this 2020 target. The Blueprint reductions can and should be supplemented by this and other sources of emissions reductions.

concludes that unchecked global warming will only create more adverse impacts on food production, public health, and species survival.

The climate will not wait for us. More recent studies have shown that the measured impacts—such as rising sea levels and shrinking summer sea ice in the Arctic—are occurring more quickly, and often more intensely, than IPCC projections (Rosenzweig et al. 2008; Rahmstorf et al. 2007; Stroeve et al. 2007).

The most expensive thing we can do is nothing. One study also estimates that if climate trends continue, the total cost of global warming in the United States could be as high as 3.6 percent of GDP by 2100 (Ackerman and Stanton 2008).

The Climate 2030 Blueprint demonstrates that we can choose to cut our carbon emissions while maintaining robust economic growth and achieving significant energy-related savings. While the Blueprint policies are not the only path forward, a near-term comprehensive suite of climate, energy, and transportation policies is essential if we are to curb global warming in an economically sound fashion. These near-term policies are also only the beginning of the journey toward achieving a clean energy economy. The nation can and must expand these and other policies beyond

Earth's climate is warming and the effects are already being felt, including more extreme heat in many parts of the country. Because carbon emissions linger in the atmosphere for 100 years or more, the U.S. must act quickly to avert the worst effects of global warming.

2030 to ensure that we meet the mid-century reductions in emissions that scientists deem necessary to avoid the worst consequences of global warming.

BOX ES.4.

Impact of the Blueprint Policies in 2020

A central insight from the Blueprint analysis is that the nation has many opportunities for making cost-effective cuts in carbon emissions in the next 10 years (through 2020). Our analysis shows that firms subject to the cap on emissions find it cost-effective to cut emissions more than required—and to bank carbon allowances for future years. Energy efficiency, renewable energy, reduced vehicle travel, and carbon offsets all contribute to these significant near-term reductions.

By 2020, we find that the United States can:
- Achieve, and go beyond, the cap requirement of a 26 percent reduction in emissions below 2005 levels, at a net annual savings of $240 billion to consumers and businesses. The reductions in excess of the cap are banked by firms for their use in later years to comply with the cap and lower costs.

- Reduce annual energy use by 17 percent compared with the Reference case levels.
- Cut the use of oil and other petroleum products by 3.4 million barrels per day compared with 2005, reducing imports to 50 percent of our needs.
- Reduce annual electricity generation by almost 20 percent compared with the Reference case while producing 10 percent of the remaining electricity with combined heat and power and 20 percent with renewable energy resources, such as wind, solar, geothermal, and bioenergy.
- Rely on complementary policies to deliver cost-effective energy efficiency, conservation, and renewable energy solutions. Excluding those energy and transportation sector policies from the Blueprint would reduce net cumulative consumer savings through 2020 from $781 billion to $602 billion.

CHAPTER 1

A Vision of a Clean Energy Economy and a Climate-Friendly Future

The writing is on the wall: the United States needs to shift away from using fossil fuels and build its economy with clean sources of energy. Many factors are driving the nation in this direction, from the need to reduce our dependence on foreign oil and head off the most devastating impacts of global warming, to calls for government investment in technologies that will spur American innovation and entrepreneurship, create jobs, and keep the United States globally competitive.

The growing threat of global warming makes this transition urgent. Global warming is caused primarily by a buildup in the atmosphere of heat-trapping emissions from human activities such as the burning of fossil fuels and clearing of forests. Oceans, forests, and land can absorb some of this carbon, but not as fast as humanity is creating it.

U.S. heat-trapping emissions have grown nearly 17 percent since 1990, with most of this increase the result of growth in CO_2 emissions from fossil fuel use in the electricity and transportation sectors. To keep the world from warming another 2°F above today's levels[10]—the level at which far more serious consequences become inevitable—the United States and other industrialized countries will have to cut emissions at least 80 percent from 2005 levels by 2050, even with swift and deep reductions by developing countries (Gupta et al. 2007; Luers et al. 2007).

We can and must accomplish this transition to a clean energy economy alongside a strong and growing U.S. economy. *Climate 2030: A National Blueprint for a Clean Energy Economy* assesses the economic and technological feasibility of meeting stringent near-term (2020) and medium-term (2030) targets for cutting global warming emissions. We analyze U.S. energy use and trends—as well as energy technologies, policy initiatives, and sources of U.S. emissions—to develop a well-reasoned, thoroughly researched, and comprehensive blueprint for action the United States can take to meet these targets cost-effectively.

1.1. The Climate 2030 Approach

Our analysis uses a modified version of the U.S. Department of Energy's National Energy Modeling System (NEMS) and supplemental analyses to conduct a comprehensive assessment of a package of climate and energy policies across multiple sectors of the economy between now and 2030. The NEMS model allows us to capture the dynamic interplay between energy use, energy prices, energy investments, the environment, and the economy, as well as the competition for limited resources under different policy scenarios.

Modeled solutions in the Climate 2030 Blueprint include more efficient buildings, industries, and vehicles; wider use of renewable energy; access to better transportation choices; and a cap-and-trade program that sets declining limits on carbon emissions.

---

10 Earth has already warmed by about 1.4°F, or 0.8°C, above the levels that existed before about 1850. An average temperature increase of 2°F above today's level is the same as a 3.6°F or 2°C increase above pre-industrial levels.

Modeled solutions include more efficient buildings, industries, and vehicles; wider use of renewable energy; and more investment in research, development, and deployment of low-carbon technologies in the electricity sector. Our model also included a cap-and-trade program that sets declining limits on emissions of carbon dioxide and other heat-trapping gases, and that makes polluters pay for "allowances" to release such emissions. A cap-and-trade program can include a provision that allows capped companies to "offset" a portion of their emissions rather than cutting them directly, by paying uncapped third parties to reduce their emissions or increase carbon storage instead. In our model, a provision for a limited amount of such offsets leads to more storage of carbon in agriculture lands and forests. (Apart from allowing for a limited number of offsets, we were unable to fully analyze the potential for storing carbon in forests and on farmland, although several studies indicate that the potential for such storage is significant [CBO 2007; Murray et al. 2005]).

Chapter 2 explains our modeling approach and major assumptions. The next four chapters then explore our major solutions in depth. Chapter 3 explains the need for an economywide price on carbon as a key driver of emissions cuts. Chapters 4–6 examine the major sectors responsible for most U.S. global warming emissions: industry and buildings, electricity, and transportation. These chapters analyze the potential savings in energy and emissions from solutions that are commercially available today, or that will very likely be available within the next two decades. The chapters also identify the challenges these solutions face in reaching widespread deployment and the policy approaches that can help overcome those challenges. (Those chapters also describe the key assumptions underlying our analysis.)

Chapter 7 presents the overall results of our analysis, while Chapter 8 provides recommendations to policy makers and other decision makers. (Our report also includes technical appendices available online, to allow readers to delve more deeply into our methods, assumptions, and results.)

1.2. Building on Previous Studies

Our analysis builds on earlier analyses of clean energy technologies and policies by university researchers, UCS, and other national nonprofit organizations over the past 15 years (Clean Energy Blueprint 2001; Energy Innovations 1997; and America's Energy Choices 1992).

Some of these reports have found that a diverse mix of energy efficiency, renewable energy, and other low-carbon technologies have the *potential* to significantly reduce heat-trapping emissions (e.g., Greenpeace International and the European Renewable Energy Council 2009, McKinsey & Company 2009, Flavin 2008, Google 2008, ASES 2007, Pacala and Socolow 2004). However, this report takes the analysis further by analyzing the costs and benefits of achieving the reductions—as well as some of the trade-offs and competition among different technologies and sectors. This report also focuses on the policy options that will enable the nation to cost-effectively meet the near-term and mid-term climate targets critical to avoiding the worst consequences of climate change.

Government agencies and university researchers have also conducted economic analyses of proposed U.S. cap-and-trade legislation (such as ACCF and NAM 2008; Banks 2008; EIA 2008; EPA 2008a; and Paltsev et al. 2007), and have analyzed the costs and benefits of implementing low-carbon technologies in specific economic sectors (such as APS 2008; EIA 2007; and EPRI 2007). However, this report again provides

FIGURE 1.1. The Sources of U.S. Heat-Trapping Emissions in 2005

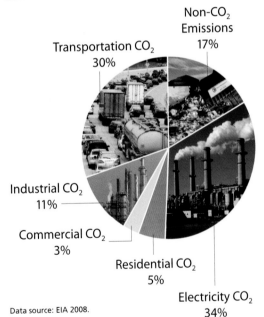

Non-CO$_2$ Emissions 17%

Transportation CO$_2$ 30%

Industrial CO$_2$ 11%

Commercial CO$_2$ 3%

Residential CO$_2$ 5%

Electricity CO$_2$ 34%

Data source: EIA 2008.

The United States was responsible for approximately 7,180 million metric tons CO$_2$ equivalent of heat-trapping emissions in 2005, the baseline year of our analysis. Most of these emissions occur when power plants burn coal or natural gas and vehicles burn gasoline or diesel. The transportation, residential, commercial, and industrial shares represent direct emissions from burning fuel, plus "upstream" emissions from producing fuel at refineries.

a more complete approach by evaluating the impact of implementing a cap-and-trade program *and* a full set of complementary energy policies and low-carbon technologies across all major sectors of the economy.

This suite of policies and technologies focuses primarily on sharply reducing U.S. emissions, with limited provisions for offsets from carbon storage in domestic lands and forests and in tropical forests. The resulting recommendations do not include every step the United States must take to address climate change. However, they establish a clear blueprint for U.S. leadership on this critical global challenge.

Addressing climate change will clearly require the participation and cooperation of both developed and developing countries. Under such a global partnership, the United States and other industrialized nations will help developing nations avoid fossil-fuel-intensive economic development and preserve carbon-storing tropical forests. The partnership will also require developed countries to fund strategies to help developing countries adapt to unavoidable climate changes.[11] Such international engagement will allow U.S companies to be at the vanguard of developing and supplying clean technologies for a global marketplace.

Although this international dimension of U.S climate policies is essential, it is beyond the scope of this report.

1.3. A Clean Energy Economy: A Solution for Many Challenges

The nation must enlist many technologies and policies if we are to meet our energy needs while addressing global warming. We propose a broad array of practical solutions to achieve our climate goals at low cost. As this report shows, many of our solutions deliver not only cost-effective cuts in global warming emissions but also consumer and business savings and other social benefits.

For example, energy efficiency technologies and measures can save households and businesses significant amounts of money. Many strategies for reducing emissions also create jobs and inject capital into the economy, while others enhance air quality, energy security, public health, international trade, and agricultural production, and help make ecosystems more resilient.

While our analysis considered most of the technologies now available to combat climate change, we focused

Tropical deforestation is one of the major causes of global warming, accounting for nearly 20 percent of global carbon emissions. The United States must therefore invest in efforts aimed at helping developing countries preserve their carbon-storing tropical forests, such as setting aside a small portion of the auction revenues from a U.S. cap-and-trade program.

on those that reduce emissions at the lowest cost, and with the fewest risks to our health and safety and the environment.

1.4. Setting a Target for U.S. Emissions Cuts

Most climate experts agree that the world must keep average temperatures from rising another 2°F above today's levels (or 2°C above pre-industrial levels) to avoid some of the most damaging effects of global warming (UCS 2008; Climate Change Research Centre 2007). Some scientists now argue that even that level is too high (Hansen et al. 2008).

In 2001 the Intergovernmental Panel on Climate Change (IPCC) identified several reasons for concern

---

11 Because global warming emissions have already accumulated in the atmosphere, the planet will undergo a certain amount of climate change regardless of future efforts to lower emissions.

regarding the world's growing vulnerability as global temperatures rise (Smith, Schellnhuber, and Qadar Mirza 2001). The arresting visual representation of this information has come to be known as the "burning embers" diagram (see Figure 1.2, left). Smith et al. (2009) drew on a 2007 IPCC report and subsequent peer-reviewed studies to update this diagram (see Figure 1.2, right).

The 2009 version highlights the much greater risk of severe impacts from rising average global temperatures than peer-reviewed studies indicated only a few years ago. The considerable evidence summarized in

---

BOX 1.1.

Causes and Effects of Global Warming

In 2007 the Intergovernmental Panel on Climate Change released a report finding that it is "unequivocal" that Earth's climate is warming, and that the planet is already feeling the effects (IPCC 2007). The primary cause of global warming is clear: burning fossil fuels such as coal, oil, and gas as we generate electricity, drive our cars, and heat our homes releases carbon dioxide and other gases that blanket the earth and trap heat. Deforestation is another major source of such emissions. To dramatically curb global warming, we will have to dramatically reduce those emissions.

Today the atmospheric concentration of two important heat-trapping gases—carbon dioxide and methane—"exceeds by far the natural range over the last 800,000 years," according to two key reports (Loulergue et al. 2008; Luthi et al. 2008). In fact, while the atmospheric concentration of heat-trapping gases was around 280 parts per million of CO_2 before 1850, it is now around 386 parts per million, and rising by almost two parts per million per year (Tans 2009).

As a result, the global average temperature is now 1.3°F (0.7°C) above pre-industrial temperatures. And the accumulation of heat-trapping gases already released ensures that the planet will warm about another 1°F (0.6°C) (Hansen et al. 2005; Meehl et al. 2005; Wigley 2005). If humanity fails to substantially reduce global emissions, the IPCC projects global average temperature increases of as much as 11.5°F (6.4°C) by the end of the century (IPCC 2007a). Such changes will likely lead to wide-ranging consequences that exceed humanity's ability to cope, including rising sea levels, widespread drought, and disruption of agriculture and global food supplies (IPCC 2007b).

Since the 2007 IPCC report, other studies have shown that climate impacts are occurring at a faster pace—and are often more intense—than IPCC projections (Rosenzweig et al. 2008; Rahmstorf et al. 2007; Stroeve et al. 2007). For example, the observed rates of both sea level rise and summer Arctic sea ice decline are higher than the IPCC anticipated in its projections.

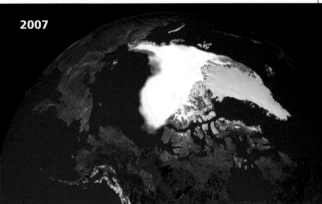

Observed and measured climate change impacts are occurring at a faster pace and are often more intense than previously projected. One example is the loss of Arctic sea ice and snow, which help reflect the sun's energy. This loss is leading to even more warming. Just 27 years after the 1980 satellite image shown here, scientists were surprised by the extent to which the minimum area of sea ice had shrunk.

FIGURE 1.2. The Risks of Climate Change: The "Burning Embers" Diagram

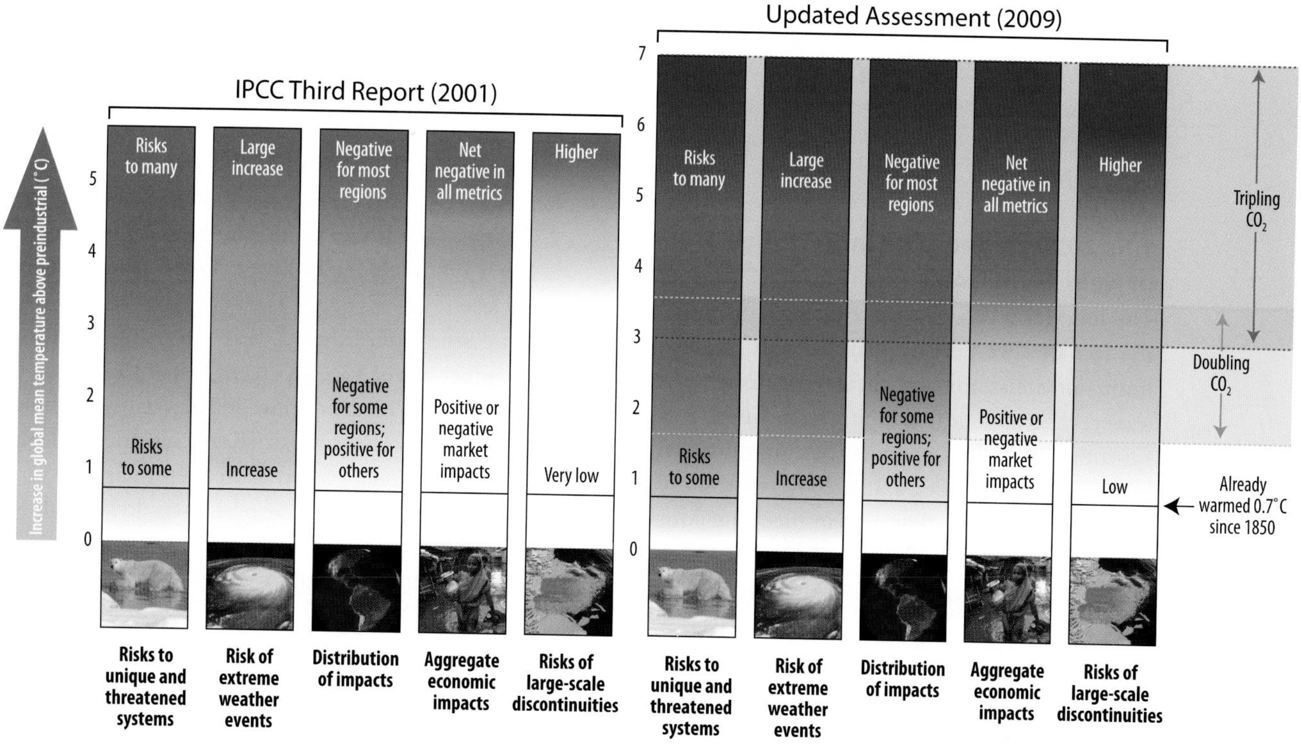

Source: Adapted from Smith et al. 2009; Schneider 2009.

The risks of harmful effects from global warming rise with its magnitude. This figure shows that even a 2°C change in global temperature poses significant risks. The left-hand panel is based on the 2001 *Third Assessment Report* (TAR) of the Intergovernmental Panel on Climate Change. The right-hand panel is an updated version from 2009.

these figures reveals that a rise in global average temperature of more than 2°F above where we are today (or 2°C above pre-industrial levels) would put many natural and human systems at grave risk.

In 2007 UCS analyzed what the United States would have to do to help keep global temperatures from rising more than 2°C above pre-industrial temperatures (Luers et al. 2007). Other studies noted that humanity has about a 50-50 chance of meeting this temperature target if we stabilize atmospheric concentrations of global warming emissions at no more than 450 parts per million of CO_2 equivalent[12] by the end of this century (Meinshausen et al. 2006). The UCS analysis therefore proposed this concentration as a maximum allowable target.

Because carbon dioxide—the primary heat-trapping gas—remains in the atmosphere for a long time, setting a target concentration also requires setting a limit for total cumulative emissions. Recent studies have shown that cumulative global emissions must not exceed about 1,700 gigatons of CO_2 equivalent[13] from 2000 to 2050, to keep atmospheric concentrations below 450 parts per million of CO_2 equivalent (van Vuuren et al. 2007; Baer and Mastrandrea 2006; Meinshausen et al. 2006).

The 2007 UCS analysis showed that the U.S. share of this budget would range from 160 to 265 gigatons CO_2 equivalent during this period, even if other nations—both industrialized and developing—acted aggressively to reduce their emissions.[14] The United States

12   Parts per million CO_2eq—a measurement that expresses the concentration of all heat-trapping gases in terms of CO_2.

13   Gigatons CO_2eq is a measure of the amount of any greenhouse gas—including CO_2 and non-CO_2 gases—based on its global warming potential compared with that of CO_2. This measure also takes into account the amount of time each gas lingers in the atmosphere. One GTCO_2eq equals 1,000 million metric tons CO_2eq.

14   The analysts developed the range for cumulative U.S. emissions by comparing the U.S. gross domestic product, population, and current emissions with those of other industrialized nations. The upper end of the range implies heroic cuts in emissions by developing countries. The prudent U.S. approach would be to stay within the mid-range of this carbon budget.

now emits about 7.1 gigatons CO_2 equivalent per year, and that amount is expected to continue to rise unless the nation establishes sound climate and energy policies. In fact, to stay within its "carbon budget," the United States would have to reduce its emissions *at least* 80 percent below 2005 levels by 2050 (Luers et al. 2007).

1.5. 2020 Targets: The Importance of Near-Term Goals

This long-term U.S. goal for reducing emissions reflects the fact that we need to plan decades in advance to limit our emissions and the severity of their consequences, because heat-trapping gases linger and accumulate over very long periods. Setting short-term and interim targets for 2020 and 2030 is therefore critical—both to ensure that we can meet our long-term

goals, and to provide the incentives and certainty that will spur firms to invest in clean energy technologies instead of locking us into high-carbon choices.

The 2007 IPCC report did not recommend specific short-term goals for cutting emissions. However, it did analyze a number of studies to determine an appropriate range of reductions for industrialized nations, to help keep global average temperatures within the 2°C target. The IPCC set this range at 25–40 percent below 1990 levels by 2020 (or 35–48 percent below 2005 levels).

One study published a year later suggested that U.S. reductions of 15–25 percent below 1990 levels by 2020 (or 27–35 percent below 2005 levels)—combined with efforts by other industrialized countries and support for developing countries to keep their emissions substantially below baseline levels—could keep global

---

BOX 1.2.

SUCCESS STORY
Reinventing Pittsburgh as a Green City

In the late 1860s, as hundreds of factories belched thick black smoke over Pittsburgh, author James Parton dubbed it "hell with the lid off" (Parton 1868). By the 1970s, as the city's industrial economy faltered, Pittsburgh's leaders made "green" buildings part of their revitalization plan. A few decades later, Pittsburgh was named the tenth-cleanest city in the world (Malone 2007).

Today Pittsburgh is a leader in green buildings, and has turned its abandoned industrial sites, known as brownfields, into assets through extensive redevelopment. Pittsburgh has shown that building green can reduce energy demand, curb global warming emissions, save consumers money on utility bills, and stimulate a green economy.

Pittsburgh's David L. Lawrence Convention Center, for example, built on a former brownfield site, is the world's first Gold LEED-certified convention center.[15] Natural daylight provides three-fourths of the lighting for the center's exhibition space, and it has reduced the use of potable water by three-fourths. Sensor-controlled lights, natural ventilation, and other efficiency measures cut energy use by 35 percent—saving the

building's owners an estimated $500,000 each year (DLCC 2009; SEA 2008).

Built on an abandoned rail yard, the PNC Firstside Center is the nation's largest Silver LEED-certified commercial building. It uses about 30 percent less energy than a traditional design, and is located near public transportation (EERE 2009). "When we see energy costs going up . . . as much as 20 percent, we think it [energy efficiency] makes fiscal sense for shareholders, employees, and the communities we do business [with]," says Gary Saulson of PNC corporate real estate (The Pittsburgh Channel 2008).

As of July 2008, Pittsburgh had at least 24 LEED-certified buildings, ranking it fifth among U.S. cities (USGBC 2008). Spurred by an initial investment from private foundations such as the Heinz Endowments and Richard King Mellon Foundation, Pittsburgh officials are now actively encouraging such efforts. In 2007, for example, the City Council adopted incentives that allow green buildings to be 20 percent taller than others in their zoning districts (City of Pittsburgh 2007). The city also created the Mayor's Green Initiative Trust Fund in 2008 with money saved through bulk power purchases

---

15 The Leadership in Energy and Environmental Design (LEED) and federal EnergyStar standards provide a framework and strategies for reducing the environmental impact of new and existing buildings, and can apply to a range of building sizes and uses.

average temperatures within the 2°C target (den Elzen et al. 2008). This analysis accepted the political reality that the United States must be allowed to start from higher baseline emissions, and set much more aggressive targets for Europe, Canada, and Russia to enable the world to remain below the maximum temperature.[16]

Another analysis, the Greenhouse Development Rights framework, considers each country's historical responsibility and current capacity to act. That framework assigns the United States responsibility for financing emissions cuts equal to 60 percent of its 1990 emission levels (or 66 percent of 2005 levels) by 2020. Some 20 percent of those cuts would come from

domestic sources, and 40 percent from efforts by other countries to reduce their emissions, funded by the United States (Baer et al. 2008).

Scientific studies alone cannot provide a specific short-term goal for cutting U.S. emissions. However, the urgency of the scientific evidence should compel the United States to set a 2020 goal that preserves our future ability to make even more aggressive reductions as we learn more about what will be necessary to stave off the worst climate impacts. **We therefore recommend that the United States reduce its global warming emissions at least 35 percent below 2005 levels (or 25 percent below 1990 levels) by 2020, primarily through domestic action.**

---

16   Having not ratified the Kyoto Treaty, the United States has experienced a steady rise in emissions since 1990.

Pittsburgh's David L. Lawrence Convention Center, which opened in 2003, uses about 35 percent less energy than a conventionally designed building of comparable size—saving the city an estimated $500,000 or more a year.

(City of Pittsburgh 2008). The fund's mandate includes the launch of a Green Council to oversee Pittsburgh's five-year plan for green initiatives.

Investing in a green economy does more than save energy: it also attracts businesses and creates jobs. The Pittsburgh region expects to see 76,000 jobs related to renewable energy during the next two decades (Global Insight 2008).That trend has already begun with the recent announcement that EverPower Wind Holdings was opening an office in the city (Schooley 2008), and with the startup of two solar manufacturing companies (Plextronics 2009; Solar Power Industries 2009).

Cities and towns play an important role in encouraging more energy-efficient buildings. Stringent energy efficiency standards for buildings, zoning incentives,

and tax rebates can encourage a clean economy. Support for targeted education and training for engineers, architects, builders, and other skilled tradespeople will ensure that the local workforce can meet growing demand for employees knowledgeable about green building.

When Pittsburgh's future seemed bleak, architect Frank Lloyd Wright was asked how to improve the city. His answer: "Abandon it!" (University of Pittsburgh 2009). Yet Pittsburgh has shown that a "green" vision, political ingenuity and persistence, and the support of private institutions can revitalize a region's economy, reduce global warming emissions, and provide a stewardship model for the nation.

CHAPTER 2
Our Approach

The Climate 2030 Blueprint provides a path for reducing U.S. heat-trapping emissions through 2030, based on scientific findings about the long-term cuts in emissions necessary to avoid the worst consequences of global warming (see Chapter 1). The long-term goals of that path are:

- To reduce annual U.S. emissions at least 80 percent from 2005 levels by 2050.
- To constrain cumulative U.S. emissions to the mid-range of a 2000–2050 U.S. "carbon budget" of 165–260 gigatons CO_2 equivalent.

The Blueprint shows how to achieve near-term and medium-term cuts in emissions through 2030 consistent with those long-term goals.

To produce the Blueprint, we considered how to curb global warming emissions from most major sources, including electricity, industry, buildings, and trans-portation, as well as some opportunities for storing carbon in U.S. forests and agricultural lands. This chapter describes how we evaluated the costs and benefits of various technologies and policies for moving along that path.

2.1. Our Model

To analyze potential cuts in U.S. emissions, we relied primarily on a modified version of the National Energy Modeling System (NEMS), developed by the Energy Information Administration (EIA), an independent division of the U.S. Department of Energy (DOE).

NEMS is a comprehensive model that forecasts U.S. energy use and emissions from the electricity, transportation, industrial, and buildings (residential and commercial) sectors. The model works by applying a variety of assumptions about technological progress and household and business behavior. It then

FIGURE 2.1. National Energy Modeling System (NEMS)

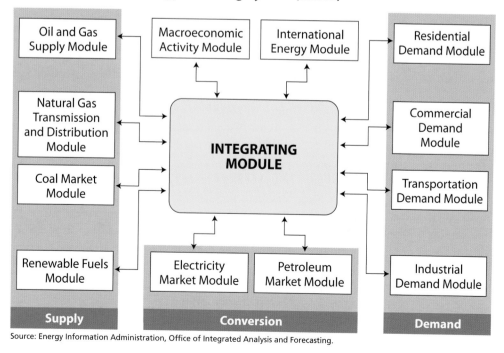

The Climate 2030 Blueprint uses a modified version of NEMS to create a forecast of the next 20 years under existing conditions. New climate, energy, and transportation policies and more advanced technologies were then added to the model to evaluate their carbon emissions-cutting potential. The Blueprint also supplemented the NEMS model with additional energy efficiency and biomass analyses.

Source: Energy Information Administration, Office of Integrated Analysis and Forecasting.

selects the technologies that can best enable the nation to meet its projected energy needs, given the assumed constraints.

The EIA uses NEMS each year to forecast U.S. energy production, demand, imports, prices, and expenditures—as well as carbon emissions—through 2030. The resulting report, known as the Annual Energy Outlook (AEO), includes a reference case based on policies in place at the time, and several "sensitivity" cases based on changes to key assumptions. The EIA also receives numerous requests from Congress to use NEMS to assess the effects of proposed climate and energy legislation.

Our approach is similar, in that we used a modified version of NEMS to create a forecast under existing policy conditions, which we call our Reference case. We then applied new climate, energy, and transportation policies to evaluate their impact in cutting heat-trapping emissions, which we call our Blueprint case.

We call our modified model UCS-NEMS (see Appendices A–G online for more on how we modified the model). We supplemented UCS-NEMS with separate analyses of the potential for making industry and buildings more efficient, and of the biomass resources potentially available to produce electricity and liquid fuel for transportation.

2.2. The Reference Case

We began our analysis with the version of NEMS used to produce *Annual Energy Outlook 2008* (EIA 2008a). The Reference case in that version of the model includes the EIA's estimates of the effects of the 2007 Energy Independence and Security Act. That law will deliver significant cuts in carbon emissions from the transportation sector by increasing fuel economy standards for cars and light trucks, and by creating a renewable fuel standard with a low-carbon requirement for most biofuels.

To establish our Reference case, we applied a variety of modifications and updates to the *AEO 2008* version of the NEMS model. For example, we modified its assumptions about the costs and performance of several energy and transportation technologies, based on data from actual projects, information from more recent studies, and input from experts. We also used the EIA's assumptions from its *AEO 2008* High Price case, which assumes higher energy prices and commodity costs. These values are more in line with the reference case forecast in *AEO 2009*, released in April 2009 (EIA 2009).

We further updated the model to include tax credits for renewable and conventional energy technologies in

Insulating homes saves energy and decreases heating and cooling bills. In the Climate 2030 Blueprint, reductions in energy use resulting from efficiency measures were calculated in a separate analysis and then modeled in NEMS to determine net energy bill savings from using less electricity and fuel in homes, businesses, and industrial facilities.

the Economic Stimulus Package signed into law in October 2008, new state renewable electricity standards, and the existing $18.5 billion nuclear loan guarantee program. However, our Reference case does not include the tax credits and incentives in the American Recovery and Reinvestment Act of 2009. (See Chapter 7 for the results of the Reference case, and the online appendices for more information on these and other modifications.)

2.3. The Climate 2030 Blueprint Case

We then developed a Blueprint case to examine the impact of bundling a cap-and-trade program with a range of complementary energy and transportation policies. We chose policies that other analyses—and real-life experience—have shown are effective in surmounting market barriers to deploying technologies that would lower energy bills and the costs of a cap-and-trade program for households and businesses (see Box 2.1). These policies stimulate improvements in energy efficiency and more widespread use of renewable energy in the industry and buildings and electricity sectors, along with cleaner cars and trucks, better transportation choices, and low-carbon fuels in the transportation sector.

We relied on an analysis by the American Council for an Energy-Efficient Economy (ACEEE) to calculate the energy and cost savings that result from the efficiency improvements in industry and buildings.[17] UCS-NEMS modeled the energy savings as drops in electricity demand and direct fuel use in industry and buildings.

The Blueprint case also included at least eight large-scale carbon-capture-and-storage (CCS) demonstration projects with advanced coal plants—consistent with the recommendation in the UCS report *Coal Power in a Warming World* (Freese, Clemmer, and Nogee 2008). Such projects can help address the technical, regulatory, and commercialization challenges of large-scale CCS technology.

The Blueprint case also accounted for existing incentives to develop and build advanced coal and nuclear plants. These include tax credits for both technologies as well as a range of risk-shifting and regulatory subsidies for nuclear plants, such as loan guarantees, insurance against licensing delays, and limits on liability.

After running the model with these modifications, we then used it to produce a "sensitivity" case—which we called the No Complementary Policies case—that stripped out the Blueprint's complementary energy and transportation policies.

2.4. The Blueprint Cap on Global Warming Emissions

A key input in the Blueprint and sensitivity cases was a trajectory for cuts in heat-trapping emissions under a cap-and-trade program beginning in 2011. We defined the U.S. cap on such emissions as a reduction of 26 percent below 2005 levels by 2020, and a drop of 56 percent below 2005 levels by 2030 (see Figure 2.2).[18]

The cumulative U.S. emissions defined by the cap—in tons of CO_2 equivalent—were direct inputs to the NEMS model (see Table 3.1 for annual values). The model then chose the most cost-effective way to comply with the cap within the constraints we imposed. Complementary policies in various sectors of the economy help deliver the cuts set by the cap more cost-effectively—they do not reduce emissions further.

Although our modeling horizon is 2030, Figure 2.2 extends the trajectory of cuts in emissions to 2050, to show that the United States will have to continue to reduce its emissions after 2030. Cuts in emissions called for by the cap accelerate each year through 2030, as the ability to manufacture and deploy low-carbon technologies grows over time. The cap continues to tighten after 2030, but at a reduced annual rate.

Other scenarios could be designed to meet the same criteria—or even show more aggressive cuts in heat-trapping emissions.

In our approach, we allowed capped firms to bank and withdraw carbon allowances (permits to emit one ton of carbon). That is, if it is cost-effective to do so, firms can choose to over-comply with the cap and bank the excess allowances for use in later years to lower emissions and costs. The result is an *actual* trajectory for emissions that differs from the trajectory specified

FIGURE 2.2. U.S. Emissions Cuts under the Blueprint Cap

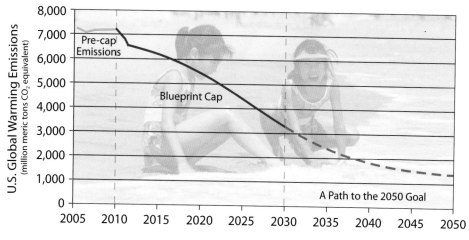

For the years 2005 to 2010, this trajectory reflects the Reference case. For the years 2011 to 2030, we modeled the impact of a cap on cumulative emissions. The trajectory beyond 2030 simply continues the deep reductions needed to stay within a long-term carbon budget, though we did not actually model what would happen from 2030 to 2050.

---

17 ACEEE conducted this analysis on our behalf. See Appendix C online for more details.

18 This differs from the recommendation in Chapter 1, which encompasses more possibilities for cutting emissions than UCS-NEMS could model, such as curbs on tropical deforestation.

by the inputs.[19] We assumed that the bank has no allowances left by 2030: that is, that there is a zero terminal bank balance. That means firms are exactly in compliance with the cap at that point.

Other studies have assumed that banks do have allowances remaining at the end of the modeling period. If the nation needs deeper cuts in emissions beyond the modeling horizon, such banked allowances could help capped firms meet their cap in those years.

For example, the EIA modeling of the cap-and-trade system in the proposed 2008 Lieberman-Warner Climate Security Act assumed an ending balance of 5 billion metric tons of CO_2 (EIA 2008). While valid, that choice is somewhat arbitrary. To accurately assess the "right" terminal bank balance, we would need precise information on the availability and cost-effectiveness of technologies for reducing emissions after 2030. In light of these uncertainties, our choice of a zero balance

19 While the model ran without constraining borrowing, the results show only banking and withdrawing. That is, the model shows no negative bank balances in any given year. This is actually the preferred policy approach because it prevents the capped sectors from delaying technological change, and also prevents the buildup of unsustainable levels of borrowing.

BOX 2.1.

Understanding Market Barriers to Climate Solutions

Many studies have documented market barriers to the development and use of cost-effective energy efficiency, conservation, and renewable energy solutions (see Chapters 4–6).

One major market failure is that energy prices do not include all the environmental, health, and national security costs of burning fossil fuels—which greater reliance on energy efficiency, conservation and renewable energy would avoid.

A second major market failure is "risk aversion": the reluctance of households and businesses to invest in climate solutions that have high up-front costs but long-term financial benefits.

A third major market failure is "split incentives" between building owners and renters. Owners do not make efficiency improvements because they do not pay the utility bills. Renters will not make the up-front investment because they are unlikely to occupy the building for long. These split incentives also exist between home developers and purchasers, and in other parts of the economy.

Other market barriers include:

- Lack of information and expertise on solutions to global warming
- Lack of capital needed for up-front investments in global warming solutions
- Lack of a core infrastructure and manufacturing capacity to support increased use of renewable energy, energy efficiency, advanced vehicle technologies, and expanded mass transit

Specific policies targeted at increasing energy efficiency, conservation, and renewable energy, such as those

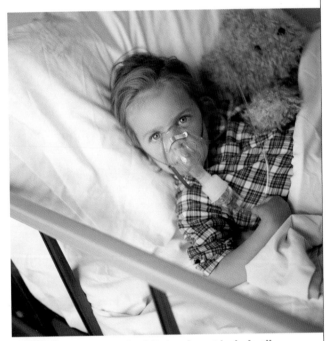

Energy prices in the United States do not include all the health, environmental, and national security costs of burning fossil fuels. Burning coal to generate electricity and gasoline in cars, for example, are significant causes of poor air quality, which in turn contributes to asthma and other respiratory problems. Increased energy efficiency and use of renewable energy can help avoid these costs.

in the Climate 2030 Blueprint, can directly address these market failures and barriers. Such policies can reduce consumers' overall costs more than energy price signals—such as those resulting from a cap-and-trade program—alone.

To determine net energy bill savings by households and businesses, the Climate 2030 Blueprint started with the combined savings on energy bills (from reduced energy use), then subtracted out the cost of purchasing more energy-efficient products such as appliances and vehicles, as well as the carbon costs passed through to consumers and businesses from their energy providers.

allows us to test the feasibility and cost-effectiveness of efforts to exactly meet the cap. However, other scenarios deserve exploration.

2.5. The Blueprint Analysis of Energy Efficiency

As noted, the Blueprint includes a supplemental analysis by the American Council for an Energy-Efficient Economy that accounts for the savings in energy and costs of nine policies and programs aimed at making industry and buildings more energy efficient.

We used the resulting national energy savings to reduce electricity demand and direct fossil fuel use in each economic sector each year. The model then distributed the energy savings across different regions of the country. In the residential sector, the model also distributed energy savings proportionally across different end-use categories, such as space heating and water heating. The model then determined the impact of the energy savings on electricity generation, the amount of fossil fuel used to produce electricity, carbon dioxide emissions, energy prices, and energy bills.

Finally, the Blueprint subtracted the investment and policy costs resulting from the efficiency analysis from savings on energy bills to determine the net savings on energy bills for consumers and businesses. (For more information, see Appendix C online.)

2.6. The Blueprint Analysis of the Biomass Supply Curve

The Blueprint also relied on a separate analysis of the amount of biomass from plant cellulose that is potentially available for use in producing electricity and liquid fuel for transportation at different prices. Marie Walsh, an agricultural economist with the University of Tennessee, and formerly of Oak Ridge National Laboratory (ORNL), conducted this analysis.

Walsh and her colleagues at ORNL developed the original supply curves used by the EIA for each of the main biomass feedstocks: energy crops (switchgrass), agricultural residues (corn stover and wheat straw), forestry residues, urban wood waste, and mill residues. The EIA model included a supply curve for each biomass feedstock for each year through 2030, and for 13 U.S. regions. The model added the data from those curves to get a total biomass supply curve for each region, and for the nation as a whole.[20]

Walsh and her colleagues at the University of Tennessee updated the supply curves for energy crops, agricultural residues, and corn for the EIA's 2007 analysis. That analysis assumed that the nation would enact policies requiring 25 percent of the electricity and energy used for transportation to come from renewable sources by 2025 (EIA 2007). The report based the supply curves on new runs of an economic forecasting model for agriculture called POLYSYS. Starting with a 2006 baseline forecast by the U.S. Department of Agriculture, the POLYSIS model projected the tonnage of all major crops and calculates changes in land use, based on the price of biomass and corn in each of 305 agricultural statistical districts.

We used those supply curves, with one exception. We reduced the amount of biomass available from energy crops by 50 percent, to account for potential indirect land-use effects that would increase carbon emissions. Such effects occur when energy crops are

---

20   The supply curve for corn is separate from that for cellulosic biomass. However, some sources of cellulosic biomass can be grown on land that could otherwise be used to grow corn, creating competition for land that can drive up the price of corn crops. UCS-NEMS does not capture this impact.

The Blueprint includes a supplemental analysis of cellulosic biomass energy that is potentially available for energy use in the United States. Biomass feedstocks are energy crops (switchgrass), agricultural residues (corn stover and wheat straw), forestry residues, and urban wood waste and mill residues.

grown on lands that could otherwise be used to grow food crops. That shift drives up the price of food crops and spurs the conversion of forests and other lands to cropland in the United States and other countries.

2.7. The Bottom Line

We chose policies for our modeling exercise with great potential to deliver critical science-based cuts in emissions cost-effectively, without other harmful consequences. We have tried to make all our assumptions transparent, so others can evaluate them on their merits. Other assumptions and combinations of technologies and policies are also possible. Analysts who investigate those alternatives should do so in a similarly transparent manner.

The rest of this chapter summarizes our major technology and policy assumptions, for quick reference. Chapters 3-6 explore these assumptions in more detail and with more context, while the appendices provide additional information. Chapter 7 presents the results of our analysis.

2.8. Summary of Blueprint Assumptions

2.8.1. Key Assumptions for the Cap-and-Trade Program

We used UCS-NEMS to model a cap-and-trade program broadly in keeping with the design criteria outlined in Chapter 3, except when constrained by specific

limitations in the model. We made the following assumptions (see Appendix B online for more details):

- The United States places a cap on global warming emissions starting in 2011. This cap declines to 26 percent below 2005 levels by 2020, and 56 percent below 2005 levels by 2030. The cap ensures that the nation is on track to stay within a mid-range carbon budget—that is, cumulative emissions—of 160–265 gigatons CO_2 equivalent from 2000 to 2050 (see Table 3.1).

- The sectors of the economy covered by the cap include electricity generation, transportation, and the industrial, commercial, and residential sectors. Household emissions from sources other than electricity are not covered.

- The cap covers emissions of all major heat-trapping gases, including CO_2 from energy production and use; CO_2 from cement and lime production; methane (CH_4) from landfills, coal mining, natural gas and oil systems, stationary and mobile combustion, and livestock; nitrous oxide (N_2O) from agriculture, stationary and mobile combustion, industrial sources, and waste management; and hydrofluorocarbons (HFCs), perfluorocarbons (PFCs), and sulfur hexafluoride (SF_6).

- Capped firms can rely on carbon "offsets" to satisfy up to 15 percent of their allowance obligations. That is, rather than cutting their emissions directly, capped companies can offset them by paying uncapped third parties to reduce their emissions or increase carbon storage. We divided the allowable offsets between domestic (a maximum of 10 percent of the cap) and international (a maximum of 5 percent of the cap).

- The federal government auctions all allowances for firms to emit carbon. However, UCS-NEMS did not allow us to channel the revenues from such auctions to investments in energy efficiency and renewable energy, or to households and businesses that may be disproportionately affected by the cap-and-trade system. We therefore simply assumed that all the proceeds from the allowance auctions would be recycled back into the economy in a general way.

- The Blueprint cap-and-trade system does not include a "safety valve"—that is, an upper limit on the price of carbon. Nor does it impose an auction reserve price, which would set a minimum price for allowances.

- Firms can bank and borrow allowances to emit carbon. We assumed that no allowances would remain in that bank in 2030. That is, the capped firms together exactly meet the target for emissions by that year.

UCS-NEMS did not allow us to model U.S. links to international cap-and-trade programs to reduce heat-trapping emissions. We were also unable to model any "leakage" of emissions: that is, undercounting of emissions stemming from imports and exports of energy-intensive goods.

2.8.2. Key Assumptions for Energy Efficiency

See the table on page 28.

2.8.3. Key Assumptions for Electricity

2.8.3.1. Key Assumptions for Technologies Used to Produce Electricity

Escalation of construction costs. We included recent increases in construction and commodity costs for all technologies, based on data from actual projects, input from experts, and power plant cost indices. We assumed that the costs of all technologies continue to rise 2.5 percent per year (after accounting for inflation) until 2015.

Wind. We included land-based, offshore, and small wind technologies. We based our capital costs on a large sample of actual projects from a database at Lawrence Berkeley National Laboratory (LBNL). We used an analysis from the National Renewable Energy Laboratory (NREL), conducted for the EIA, to develop regional wind supply curves that include added costs for siting, transmitting, and integrating wind power as its use grows.

We also assumed increases in wind capacity factors (a measure of power production) and a 10 percent reduction in capital costs by 2030 from technological learning, based on assumptions from a report from the DOE on producing 20 percent of U.S. electricity from wind power by 2030 (EERE 2008).

Solar. We assumed expanded use of concentrating solar power (CSP) and distributed (small-scale) and utility-scale photovoltaics through 2020, based on actual proposals. We also assumed faster learning for solar photovoltaics, to match the EIA's assumptions for other emerging technologies. We assumed that the amount of heat that CSP can store to produce electricity during periods of high demand rises over time.

Bioenergy. Key technologies included burning biomass along with coal in existing coal plants, dedicated biomass gasification plants, the use of biomass to produce combined heat and power in the industrial sector, and the use of methane gas from landfills.

Geothermal. We included a supply curve for hydrothermal and enhanced geothermal systems in the West, developed by NREL and other experts. This supply curve incorporates recent increases in the costs of exploring potential sites, drilling, and building geothermal power plants.

Hydropower. We assumed incremental amounts of hydropower from upgrades and new capacity at

Farmers and property owners who lease their land for wind farms—such as these landowners in Somerset, PA—enjoy a steady stream of extra income, while nearby towns and communities benefit from a larger tax base.

The Searsburg, VT, wind project reduces CO_2 emissions in New England by more than 6,600 tons annually—the equivalent of taking over 900 cars off the road.

Key Policies for Improving the Energy Efficiency of Industry and Buildings	Total Savings in 2030 (in End-Use Quads)	Total Cost in 2030 (in Billions of 2006 Dollars)	
		Program	**Investment**
Appliance and equipment standards: The federal government upgrades energy efficiency standards or establishes new ones for 15 types of appliances and equipment over the next several years.	1.8	0.50	11.45
Energy efficiency resource standard (EERS): Federal standards rise steadily to 20 percent for electricity and 10 percent for natural gas by 2030.	3.7	1.63	16.26
Building energy codes: New codes cut energy use in new residential and commercial buildings 15 percent annually until 2020, and 20 percent annually from 2021 to 2030.	1.2	2.12	14.19
Advanced buildings: An aggressive program ramps up and results in an additional 15 percent drop in energy use in new residential and commercial buildings by 2023 (beyond minimum building codes), with savings continuing at that level through 2030.	1.1	3.96	21.78
Research and development: Annual R&D investments reach $4.6 billion in 2030, and stimulate additional private-sector investments that reach $18.5 billion that year. These investments result in a 4.4 percent reduction in U.S. energy use by 2030.	1.8	4.65	18.50
Combined heat and power (CHP): A range of barrier-removing policies and annual investments in federal and state CHP programs lead to about 88,000 megawatts of new capacity by 2030—an average annual addition of 4,000 megawatts.	0.6	0.06	27.57
Industrial energy efficiency: Expanded federal programs, combined with local programs that support plant-level efforts, reduce industrial fuel use 10 percent (beyond that achieved by EERS and CHP) by 2030.	1.7	0.36	2.58
Rural energy efficiency: The federal government expands its Farm Bill Section 9006 technical assistance grants.	0.01	0.003	0.02
Petroleum feedstocks: Wider use of recycled feedstocks cuts industrial use of petroleum feedstocks 20 percent by 2030.	0.3	0.02	0.15
TOTAL	12.1	13.40	113.55

existing dams, and counted both new sources of power as contributing to a national standard for renewable electricity.

Carbon capture and storage. We included this as an option for advanced coal gasification and natural gas combined-cycle plants, with costs and performance based on recent studies and proposed projects.

Nuclear. We assumed that existing plants are relicensed and continue to operate through their 20-year license extension, and that they are then retired, as the EIA also assumes. We based assumptions on the costs and performance of new advanced plants primarily on recent project proposals and studies.

Transmission. We included the costs of new capacity for transmitting electricity for all renewable, fossil fuel, and nuclear technologies. We also added costs for the growing amounts of wind power, based on the NREL analysis conducted for the EIA.

See Appendix D online for more details.

2.8.3.2. Key Assumptions for Electricity Policies

Policies in the Reference case include:

State renewable electricity standards. These specify the amount of electricity that power suppliers must obtain from renewable energy sources. We replaced the EIA's estimate with our own projections for state standards through 2030. We applied those projections to the 28 states—plus Washington, DC—with such standards as of November 2008.

Tax credits. We included the tax credit extensions for renewable energy and advanced fossil fuel technologies that were part of the Economic Stimulus Package (H.R. 6049) passed by Congress in October 2008.

Nuclear loan guarantees. We assumed that the $18.5 billion in loan guarantees spur the construction of four new nuclear plants with 4,400 megawatts of capacity by 2020, based on applications received by the U.S. Department of Energy in October 2008.

Additional policies in the Blueprint include:

Efficiency. Policies to increase energy efficiency in buildings and industry (see Chapter 4) reduce electricity demand 35 percent by 2030, compared with the Reference case.

Combined heat and power (CHP). Policies and incentives to increase the use of natural gas combined-heat-and-power systems in industry and commercial buildings (see Chapter 4) enable this technology to provide 16 percent of U.S. electricity generation by 2030.

National renewable electricity standard. This standard requires retail electricity providers to obtain 40 percent of remaining electricity demand (after reductions for efficiency improvements and CHP) from renewable energy (wind, solar, geothermal, bioenergy, and incremental hydropower) by 2030.

Coal with carbon capture and storage (CCS) demonstration program. This new federal program provides $9 billion to cover the incremental costs of adding CCS at eight new full-scale advanced coal plants—known as integrated gasification combined-cycle plants, which turn coal into gas—from 2013 to 2016 in several regions.

Transmission. We included the costs of new capacity for transmitting electricity for all renewable, fossil fuel, and nuclear technologies. We also added costs for the growing amounts of wind power, based on the NREL analysis conducted for the EIA.

A combined-heat-and-power plant at the University of New Hampshire provides heating, cooling, and electricity to the majority of buildings on the main campus while reducing carbon emissions and saving money.

2.8.4. Key Assumptions for Transportation Technology

Cars and light trucks. We based the cost and performance of technology for improving the fuel economy of cars and light trucks on the NEMS advanced technology case. That case is slightly more pessimistic than UCS estimates, but it is more optimistic and includes more technology options than estimates by the National Research Council (NRC 2002).

Medium- and heavy-duty truck technology. We based the cost and performance of these vehicles on the NEMS advanced technology case. However, we modified the cost and performance of hybrids, and technologies to improve vehicle and trailer aerodynamics, based on discussions with the authors of a forthcoming report on the fuel-economy potential of heavy-duty vehicles (Cooper et al. forthcoming).

Moving containerized freight long distances by rail is far more energy-efficient than moving those goods by truck, releasing fewer carbon emissions and alleviating truck traffic on our nation's highways.

Vehicle air conditioning. We based the cost and performance of improved air conditioning on information from the California Air Resources Board (CARB 2008), and on a UCS research report (Bedsworth 2004). The latter assumed that manufacturers switch to HFC-152a, though a switch to HFO1234yf could provide even greater reductions in heat-trapping emissions.

Biofuels. Key technologies include ethanol from plant cellulose and biomass-to-liquid gasification technology. We initially based the costs for both on the NEMS High Commodity Cost case. However, we then reduced the cost of cellulosic ethanol by 38 percent, based on data from NREL. Biomass resources include crop residues and dedicated energy crops such as switchgrass. We excluded forest, urban, and mill residues because of limitations in the NEMS model. We also excluded 50 percent of crop-based resources, to reflect sustainability criteria and minimize indirect effects on land use.

Refineries. We assumed a 10 percent increase in refinery efficiency, based on an assessment of existing potential by analysts from the LBNL.

Advanced vehicles. The portfolio of potential advanced vehicles includes plug-in hybrids, battery-electric vehicles, and fuel cell vehicles. We used plug-ins as the sole technology for ease of modeling, rather than applying a performance-based requirement for advanced vehicles. However, other technologies with equal performance could substitute. We drew information on the cost and performance of plug-ins from research at MIT (Bandivadekar et al. 2008).

Transit. We based the costs of doubling the amount of public transit nationwide on estimates from the American Association of State Highway and Transportation Officials (AASHTO).

Pay as you drive. We based the driver response to pay-per-mile policies on studies from Cambridge Systematics (Cowart 2008) and the Brookings Institution (Bordoff and Noel 2008). We used the analysis from the Brookings Institution to determine the costs of GPS-based odometer tracking.

Fuel Economy Potential and Costs Used in the Climate 2030 Blueprint	Cars and Light-Duty Trucks	Medium- Duty Trucks	Heavy- Duty Trucks
2005 Baseline Fuel Economy (mi/gallon gasoline eq)	26	8.6	6
2020 Fuel Economy for New Vehicles (mi/gallon gasoline eq)	42	11	8
2020 Incremental Cost vs. 2005 (2006 dollars)	$2,900	$6,000	$15,800
2030 Fuel Economy for New Vehicles (mi/gallon gasoline eq)	55	16	9.5
2030 Incremental Cost vs. 2005 (2006 dollars)	$5,200	$14,900	$40,500

Notes: These potentials and costs are based on assumptions in the AEO 2008 NEMS high technology case, as modified by the authors, and modeling runs of UCS-NEMS. The values in our Blueprint case model runs may not match these levels because of limitations in the model. See Appendix E online for details.

Standards for Vehicle Global Warming Emissions	Cars and Light-Duty Trucks	Medium- Duty Trucks	Heavy- Duty Trucks
2005 Baseline Global Warming Emissions (g/mi CO$_2$eq)[a]	**372**	**1,038**	**1,489**
Fuel Economy (mi/gallon gasoline eq)	24	8.6	6
Non-CO$_2$ Emissions Estimate (g/mi CO$_2$eq)	2	5	8
2020 Standard for Global Warming Emissions (g/mi CO$_2$eq)[a]	**198**	**777**	**1,072**
Fuel Economy (mi/gallon gasoline eq)	42	11	8
CO$_2$ Emissions with Current Gasoline (g/mi CO$_2$eq)[b]	212	808	1,111
Non-CO$_2$ Emissions Estimate (g/mi CO$_2$eq)[c]	2	5	8
Credit for Improved A/C (g/mi CO$_2$eq)[d]	-8	-8	-8
Credit for Low-Carbon Fuel Standard (g/mi CO$_2$eq)[e]	-7	-28	-39
2030 Standard for Global Warming Emissions (g/mi CO$_2$eq)[a]	**139**	**497**	**842**
Fuel Economy (mi/gallon gasoline eq)	55	16	9.5
CO$_2$ Emissions with Current Gasoline (g/mi CO$_2$eq)[b]	162	555	935
Non-CO$_2$ Emissions Estimate (g/mi CO$_2$eq)[c]	2	5	8
Credit for Improved A/C (g/mi CO$_2$eq)[d]	-8	-8	-8
Credit for Low-Carbon Fuel Standard (g/mi CO$_2$eq)[e]	-16	-56	-94

Note: Values may not sum properly because of rounding.

a We calculated global warming emissions as the sum of CO$_2$ and non-CO$_2$ emissions from today's gasoline, minus cuts in emissions from the use of better air conditioning and low-carbon fuels.

b In converting fuel economy into CO$_2$ equivalent, we assumed 8,887 grams of CO$_2$ per gallon of today's gasoline burned.

c We scaled up estimates of non-CO$_2$ heat-trapping emissions for medium- and heavy-duty trucks from those for light-duty vehicles based on relative fuel consumption. We expect to update these numbers as more accurate data become available. These estimates do not include black carbon.

d Note that 8 grams per mile is a conservative estimate for cars and light trucks based on Bedsworth 2004 and CARB 2008. We have no data for medium- and heavy-duty vehicles. However, given that they have larger air conditioning systems (and thus greater potential for absolute savings) but travel farther (reducing the per-mile benefit), we used 8 grams per mile as a rough value pending more information.

e All fuels achieve the average low-carbon standard in Table 6.4.

A Look at Cellulosic Ethanol in 2030			
Resource, Yield and Potential		**Costs**	
Biomass Resources Available for Transportation (million tons)	280	Fixed Production Costs (in 2006 dollars per gallon)	$0.128
Ethanol Yield (gallons per ton)	110	Non-Feedstock Variable Costs (in 2006 dollars per gallon)	$0.17
Maximum Biofuel Potential (billion gallons ethanol equivalent)	30	Initial Capital Cost (in 2006 dollars per gallon of capacity)	$1.99

Note: In our Blueprint analysis, actual production of cellulosic ethanol may be lower, as it competes with biomass-to-liquids technology for access to biomass resources. However, the total volume of low-carbon biofuels will be similar.

Potential of Advanced Vehicles and Fuels	2020	2030
Low-Carbon Fuel Standard: Reduction in Carbon Intensity for All Transportation Fuels vs. 2005[a]	3.5%	10%
Sales of Advanced Light-Duty Vehicles Spurred by Regulations[b]	2.0%	20%

Notes:

a   This standard would require a reduction in life-cycle grams of CO_2 equivalent per BTU of all fuel used for transportation, including cars and light trucks, medium- and heavy-duty vehicles, rail, air, shipping, and other miscellaneous uses. If the standard is restricted to highway vehicles (cars, light trucks, and medium- and heavy-duty vehicles), the figure for 2020 would be 4.5 percent, and that for 2030 would be 14 percent.

b   This represents the fraction of light-duty vehicles that are plug-in hybrids, or pure battery and fuel cell vehicles delivering equivalent benefits.

Potential for Reducing Vehicle Miles Traveled	2020	2030
Assumed Policy Impact: Reduction in Annual Growth in Vehicle Miles Traveled (VMT)[a]		
Light-Duty Vehicles[b]	Reduce growth in VMT from baseline of 1.4% per year to 0.9% per year	
Trucks[c]	Reduce VMT by 0.1% per year, on top of all other policy effects	
Policies and Costs for Light-Duty Vehicles		
Transit[d]	Ramp up transit funding to reach $21 billion per year by 2030	
Pay as You Drive		
Highway User Fee 1: Maintain Existing Funding Levels[e]	$0.005 per mile	$0.011 per mile
Highway User Fee 2: Congestion Mitigation Fee Used to Fund Transit[d]	$0.004	$0.006
Total User Fees	$0.009 per mile	$0.017 per mile
Pay-as-You-Drive (PAYD) Insurance[e]	$0.07 per mile	$0.07 per mile
Federal Funding for PAYD Pilot Programs	$3 million per year for 5 years	
Tax Credit for PAYD Electronics	$100 million per year for 5 years	
Smart Growth[f]	$0.00	$0.00
Policies and Costs for Heavy-Duty Vehicles		
Switch from Truck to Rail[g]	$0.00	$0.00

Notes:

a   NEMS is unable to model the full suite of policies needed to address vehicle travel. Instead, we inserted the total reductions in vehicle miles traveled that could result from such policies into UCS-NEMS.

b   For the potential to reduce VMT from light-duty vehicles, we relied primarily on a recent analysis by Cambridge Systematics (Cowart 2008), which found that growth in light-duty VMT could be reduced to 0.9 percent per year.

c   To evaluate the potential to reduce VMT from freight trucks, we assumed that policies can shift 2.5 percent of truck VMT to rail, based on potential highlighted in AASHTO 2007 and IWG 2000. This represents about a 0.1 percent annual reduction in freight truck travel. Actual freight truck travel will fall further as the economy shifts due to other policies, such as a cap-and-trade program and reduced oil use from higher vehicle efficiency.

d   The congestion mitigation fee provides this funding, so we did not count it as a cost above that fee.

e   Blueprint policies do not include these fees as a cost, because the Reference case would also need to raise the highway funding to pay for repair of existing roads, and would include the cost of insurance. Actual insurance costs would probably drop, because people would drive less under the Blueprint.

f   Smart-growth policies could actually reduce costs, so we assumed that they are cost-neutral.

g   Switching from truck to rail will likely entail some costs, but evaluating them was beyond the scope of our study.

NUMBER OF 100°F DAYS PER YEAR IN CHICAGO: PROJECTIONS FROM THE CHICAGO CLIMATE ACTION PLAN

Higher-Emissions Scenario
31 days

16 days

7 days

Lower-Emissions Scenario
8 days

5 days

2 days

1961–1990   2010–2039   2040–2069   2070–20

Higher-Emissions Scenario   Lower-Emissions Scenario

CHAPTER 3

Putting a Price on Global Warming Emissions

We know that global warming poses a grave risk to our planet and our very way of life. As Chapter 1 pointed out, to avoid some of the worst consequences, we must keep the global average temperature from rising more than 2°F from today's levels. To meet this goal, the United States will have to cut its emissions at least 80 percent from 2005 levels by 2050, and keep its cumulative emissions from 2000 to 2050 at 160–265 gigatons CO_2 equivalent (Luers et al. 2007).

A well-designed economywide cap-and-trade program that sets a declining limit on heat-trapping emissions while charging polluters for their emissions is a lynchpin of a comprehensive approach to addressing global warming. The nation will also need to pursue other measures, including sector-based policies that promote energy efficiency, the use of renewable energy, and better transportation and fuels (see Chapters 4–6). Investments in reducing emissions in other nations are also essential, such as funding for protection of tropical forests and deployment of clean technology in developing countries.

3.1. How a Well-Designed Cap-and-Trade Program Works

One primary goal of a cap-and-trade program is to put a price on heat-trapping emissions and require polluters to pay for their pollution. Such a system will encourage the entire economy to look for cost-effective ways to reduce these emissions, and help usher in clean technologies and innovation essential to making the transition to a carbon-free economy.

The "cap" refers to the strict, declining limit on economywide heat-trapping emissions that the nation must set to help avoid the worst effects of global warming. Federal legislation should establish this cap, but it should also be adjustable over time to reflect the latest scientific information.

Under the program, regulated companies (the emitters) would have to purchase permits—also called allowances—for all their heat-trapping emissions. The total number of allowances issued would match the level of emissions allowed under the cap. Allowances would be made available through regularly held auctions.

With this basic framework, a cap-and-trade program creates a market for emissions allowances and spurs companies to curtail their emissions by financially rewarding more climate-friendly practices. For example, power producers may choose to shift from fossil-fuel-intensive sources of electricity such as coal to renewable sources such as wind and solar energy. Entrepreneurs who develop and sell new low-carbon technologies, such as improved solar panels or techniques for storing carbon in soils and trees, will also see a robust market for their products and respond accordingly. Households and businesses may also respond by, for example, purchasing more efficient appliances or equipment to reduce their energy costs. Together these actions will help the nation achieve the targeted cuts in emissions at the lowest cost.

In the jargon of economics, the "price signal"—that is, the price of the allowances, as set through the auction—helps correct (or "internalize") a market failure (or "externality") that has allowed companies to make

LEFT: One serious consequence of unchecked climate change will be extreme heat in cities. For example, under a high-emissions scenario (the red line), Chicago may experience 31 days above 100°F every year by the end of this century (City of Chicago 2008). Actions taken today—such as those called for in the Climate 2030 Blueprint—can help prevent such extreme heat. (Note: The Blueprint analysis is based on a different emissions scenario and extends only to 2030; see Table 3.1).

World-renowned economist Sir Nicholas Stern has stated that, "Climate change is the greatest market failure the world has ever seen." A well-designed cap-and-trade program can help address this failure in a cost-effective way while also providing an incentive for developing and deploying clean technologies throughout the economy.

production and investment decisions, and households to make consumption choices, without accounting for the societal costs of the resulting pollution. In the words of Sir Nicholas Stern, author of the 2006 *Stern Review of the Economics of Climate Change*, "Climate Change is the greatest market failure the world has ever seen." A well-designed cap-and-trade program can help address this market failure in a cost-effective way that benefits the overall society.

An alternative approach to putting a price on emissions is a carbon tax. In theory, a carbon tax and a cap-and-trade program have equivalent effects. (One sets a fixed price that then determines the quantity of emissions, while the other sets a fixed quantity of emissions that determines their price.)

However, the one fundamental problem with a tax is that it is impossible to know ahead of time what level the tax should be to produce the cuts in emissions we need. Policy makers would also be unlikely to continuously adjust the tax to meet a specific target for emissions. The price of allowances in a cap-and-trade program adjusts automatically, in contrast, to account for changing market conditions,[21] but always ensuring the necessary emissions cuts are achieved.

A cap-and-trade program may have greater merit as a practical matter of policy. Both approaches could also coexist. For example, policy makers could impose a carbon tax on sectors that are hard to include under a cap. However, it is critical that a well-designed market instrument be put in place without delay, to jump-start our transition to a clean energy economy.

3.2. A Tried-and-Tested Approach

The United States pioneered the cap-and-trade approach—to control emissions of sulfur dioxide (SO_2), a major component of acid rain, which acidifies lakes and forests and poses threats to public health. These emissions are primarily a by-product of burning coal to produce electricity.

The Acid Rain Program, which created an SO_2 trading system, was part of the Clean Air Act of 1990.[22] The program required owners of coal-burning power plants to reduce their SO_2 emissions to 50 percent of their 1980 levels by 2010.[23] The Acid Rain Progress Report from the U.S. Environmental Protection Agency (EPA) shows that the nation reached this goal in 2007—three years before the statutory deadline—and at only one-fourth the estimated cost (EPA

---

21 For example, current prices of allowances under the European Union's cap-and-trade program are low because the recession has reduced demand for energy, and thus the need for allowances (see Box 3.2).

22 The Acid Rain Program also limited emissions of nitrous oxide (NO_X), another contributor to acid rain. The program achieved these cuts through more traditional regulatory means.

23 However, as scientists have tracked the impact of the remaining SO_2 emissions on our nation's lakes and forests, they now realize that further reductions will be needed. In other words, regulators may have to lower the cap to account for new scientific information.

2007).[24] The report estimated that the public health benefits of the program exceeded its costs by more than 40:1.

Drawing on this experience, Europe, Australia, and New Zealand—as well as several U.S. states and regions—have committed to or already implemented cap-and-trade programs for heat-trapping emissions. As this approach becomes the international market tool of choice, the United States must place an even higher priority on developing a sound economywide cap-and-trade program, and ultimately link that program with those created by other nations. (See Box 3.2 for lessons from existing and proposed cap-and-trade programs.)

3.3. Design for Success

A successful cap-and-trade program must be designed well from the outset. Several critical features will help make it robust, transparent, fair, and effective:

Setting a stringent, declining cap on heat-trapping emissions, with firm near-term and long-term goals and a tight budget for cumulative emissions.

As noted, the United States must reduce its heat-trapping emissions at least 80 percent below 2005 levels by 2050, to avoid the worst effects of global warming. Delay in taking action will require much sharper cuts later, which would likely be more difficult and costly.

To start the nation on the path to the 2050 target, climate policies should require at least a 35 percent drop from 2005 levels by 2020, primarily from U.S. sources, and also from investments in cutting emissions in other countries. Thus the cap-and-trade program should set a cap on U.S. emissions to match this level of ambition. Because our understanding of climate science advances continuously, the program should also require regular reviews of the latest information, and include a mechanism for adjusting the target for emissions if needed.

However, these percentage reductions do not tell the whole story, because heat-trapping emissions accumulate and persist in the atmosphere for long periods of time (more than 200 years, in the case of CO_2). Thus the critical metric of success of a cap-and-trade program is a stringent budget for cumulative carbon emissions. Chapter 1 suggests a U.S. budget of 165–260 gigatons CO_2 equivalent from 2000 to 2050. Of

Acid rain, which is caused by sulfur dioxide emitted from coal-burning power plants, contributes to human health problems and can kill aquatic plants and animals and destroy forests (as shown here). The world's first cap-and-trade system was established in 1990 as part of a U.S. effort to address acid rain, and it has proven a success, achieving its 2010 emissions reduction goal three years ahead of schedule, at a much lower cost than originally expected.

---

24   A primary reason for these lower compliance costs was the switch to low-sulfur coal from Wyoming—made cheaper by the deregulation of railroad freight. Other reasons include more output from nuclear power plants as a result of higher "capacity factors" (the ability to run at full capacity during more hours of the year); a decline in natural gas prices, coupled with efficiency improvements in natural gas combined-cycle plants, which led to greater reliance on those plants; and technological innovations that led to lower-cost, better-performing scrubbers, which reduce sulfur emissions 90 percent or more.

this amount, the nation will already have emitted about 78 gigatons by 2010, at today's rate of about 7.1 gigatons a year.

Including as many economic sectors as possible under the cap.

The cap should cover all major sources of emissions—either directly or indirectly—to ensure that the needed economywide reductions occur, and to spur all sectors to adapt their production and investments in response to the price of emissions. A cap-and-trade program should also provide incentives for sources that may remain uncapped (such as the agriculture and forestry sectors) to reduce their emissions, by using the proceeds from auctioning allowances to set standards and fund programs.

Including all major heat-trapping emissions.

To exert the greatest impact, the cap should apply to all major heat-trapping emissions, including—but not limited to—carbon dioxide (CO_2), methane (CH_4), nitrous oxide (N_2O), hydrofluorocarbons (HFCs), perfluorocarbons (PFCs), and sulfur hexafluoride (SF_6).

Auctioning all allowances rather than giving them away free to emitters, and using the revenues to advance the public good.

An auction would be the most efficient and equitable way to distribute allowances to release emissions. While firms would bear the regulatory burden of purchasing the allowances, they would not necessarily pay the final costs. Most companies would pass on these costs to consumers, regardless of whether the program auctioned the allowances or gave them away.

Once carbon emissions are capped, allowances become a valuable commodity. Giving them away for free would most likely result in windfall profits for companies without producing any benefit for consumers.[25] Instead, the government should auction the allowances and use the revenues for productive purposes—an approach known as "revenue recycling." The government could direct some of these revenues to consumers, to offset the costs that companies pass through to them. Recent studies by the Congressional Budget Office (CBO 2007a) and Dinan and Rogers (2002) have documented the economic benefits of an allowance auction with revenue recycling.

Ten northeastern states already have a functioning cap-and-trade system to reduce global warming emissions. The states auction nearly all of their emissions permits (or allowances) and invest the revenue in clean energy technologies and policies. Governor John Lynch of New Hampshire declared, "We need to continue to invest in energy efficiency and work to ensure that 25 percent of our energy comes from renewable power by 2025.... My Green Jobs Initiative will help create jobs for our people now and make New Hampshire's economy stronger for the future."

Auction revenues from a stringent cap-and-trade program will total hundreds of billions of dollars per year. The government can invest these funds in measures that promote cuts in emissions, such as clean, renewable energy technologies, energy efficiency, and efforts to protect forests in developing countries.

Government should also invest the funds in measures that help consumers and communities transition to a low-carbon economy. These measures include rebates for low-income families, transition assistance to workers who are disproportionately affected by the program, and help for communities and ecosystems in adapting to the unavoidable effects of global warming.

Excluding loopholes that undermine the program.

A cap-and-trade program should not include a "safety valve" that short-circuits the market by setting a maximum price on allowances—above which an unlimited number would become available. This approach would distort the market, undermine the nation's ability to fulfill its goals for cutting emissions, and reduce the incentive for companies to invest in developing and using clean technologies.

---

25   This is a critical lesson from experience with the European Union's cap-and-trade program (see Box 3.2).

Limiting "offsets."

Rather than reducing their own emissions or buying allowances, regulated companies may purchase "offsets" by paying parties or countries not subject to the cap to reduce their emissions. For example, a power producer could offset its emissions by paying an unregulated landfill owner to capture methane emissions.

If these offsets are cheaper than efforts to cut emissions directly, they can help polluters lower the costs of complying with a cap. Offsets also allow unregulated entities and countries to contribute to the global effort to reduce heat-trapping emissions.

However, by helping major emitting sectors of the economy postpone cuts in their own emissions, offsets could delay the much-needed technological transformation and innovations in those capped sectors. That, in turn, would jeopardize the cap-and-trade program's long-term goals, and perhaps raise its long-term costs. Ensuring that offsets meet stringent criteria—such as that they are real, verifiable, quantifiable, additional (that is, beyond any that would have occurred without the program), permanent, and enforceable—may also require considerable resources.

> **Banking allows firms to choose which technologies to invest in and make other investment decisions over a longer time frame, and thus greatly reduces the volatility of the price of emissions allowances.**

The nation may find cheaper and more efficient ways to spur cuts in emissions from sectors that are not easy to cap. These could include direct mandates (such as performance or technology standards), financial incentives funded by auction revenues, subsidies, and loan guarantees.

An effective cap-and-trade program should therefore limit the number of offsets that capped companies can rely on. Any offsets that the program does allow should meet strict quality standards, and it should include a strong institutional framework for monitoring and enforcing those standards. The total number of offsets allowed must relate directly to the stringency of the cap. For example, regulators could mandate that companies could use offsets to meet only a small percentage of their required cuts in emissions.

Allowing banking and borrowing.

Banking would allow firms to exceed their required cuts in emissions in early years and store up credits for use in later years. Borrowing would allow firms to emit more global warming pollution early if they commit to making sharper cuts in emissions later.

Banking allows firms to choose which technologies to invest in and make other investment decisions over a longer time frame, and thus greatly reduces the volatility of the price of emissions allowances. Unrestricted banking will spur early cuts in emissions, which are important for safeguarding the climate.

However, as with offsets, early borrowing at unsustainable levels can lead capped sectors to postpone cuts in emissions, and can undermine the program's overall goals. Policy makers should therefore limit the amount of borrowing firms can do, such as by imposing a three-year "true-up" period, so borrowing cannot get out of hand.

Creating strong institutions.

A cap-and-trade program requires a strong institutional framework to function well. The EPA—the agency that would oversee the program—will play a critical role in ensuring that it achieves its goals. The EPA will have to work closely with scientists, policy makers, and the authority that will oversee the market for trading allowances.

That authority, in turn, will have to guard against "gaming" or other illegal activities that interfere with the proper functioning of allowance auctions. It will also have to oversee any secondary markets for trading allowances that will develop as capped firms and other parties (including brokers and investors) trade allowances.

The EPA must have enough resources to ensure that regulated companies comply with their requirements, and that they face appropriate penalties if they do not. The agency will also have to strictly monitor and enforce standards for offsets.

Meanwhile a trustworthy fiduciary entity must oversee the disbursement of revenues from the sale of allowances, to ensure that they go to the appropriate recipients for the appropriate purposes. Congress or the EPA will also have to choose an authority to manage links between a domestic cap-and-trade program and international carbon markets.

Finally, a robust, high-quality cap-and-trade program needs excellent baseline data on emissions, and the ability to track them over time.

Linking with similar programs.

Linking a U.S. cap-and-trade regime with well-designed cap-and-trade programs in other regions can provide important economic advantages, such as enabling capped companies to find the lowest-cost sources of reductions over a wider geographic area. Such links would require that the regimes be compatible, especially with regard to the stringency of the cuts in emissions they require and other key program standards.

The NEMS model looks for the most cost-effective way of meeting the cumulative goal for emissions (in tons) over the entire modeling period of 2011 to 2030, taking into account banking and borrowing of allowances. This means that the actual year-by-year emissions that are shown in the model results (the "actual emissions trajectory") may differ considerably from

---

BOX 3.1.

Climate 2030 Blueprint Modeling Assumptions: Cap-and-Trade Program

We used UCS-NEMS to model a cap-and-trade program broadly in keeping with the design criteria outlined in Chapter 3, except when constrained by specific limitations in the model. We made the following assumptions (see Appendix B online for more details):

- The United States places a cap on global warming emissions starting in 2011. This cap declines to 26 percent below 2005 levels by 2020, and 56 percent below 2005 levels by 2030. The cap ensures that the nation is on track to stay within a mid-range carbon budget—that is, cumulative emissions—of 160–265 gigatons CO_2 equivalent from 2000 to 2050 (see Table 3.1).

- The sectors of the economy covered by the cap include electricity generation, transportation, and the industrial, commercial, and residential sectors. Household emissions from sources other than electricity are not covered.

- The cap covers emissions of all major heat-trapping gases, including CO_2 from energy production and use; CO_2 from cement and lime production; methane (CH_4) from landfills, coal mining, natural gas and oil systems, stationary and mobile combustion, and livestock; nitrous oxide (N_2O) from agriculture, stationary and mobile combustion, industrial sources, and waste management; and hydrofluorocarbons (HFCs), perfluorocarbons (PFCs), and sulfur hexafluoride (SF_6).

- Capped firms can rely on carbon "offsets" to satisfy up to 15 percent of their allowance obligations. That is, rather than cutting their emissions directly, capped companies can offset them by paying uncapped third parties to reduce their emissions or increase carbon storage. We divided the allowable offsets between domestic (a maximum of 10 percent of the cap) and international (a maximum of 5 percent of the cap).

- The federal government auctions all allowances for firms to emit carbon. However, UCS-NEMS did not allow us to channel the revenues from such auctions to investments in energy efficiency and renewable energy, or to households and businesses that may be disproportionately affected by the cap-and-trade system. We therefore simply assumed that all proceeds from the allowance auctions would be recycled back into the economy in a general way.

- The Blueprint cap-and-trade system does not include a "safety valve"—that is, an upper limit on the price of carbon. Nor does it impose an auction reserve price, which would set a minimum price for allowances.

- Firms can bank and borrow allowances to emit carbon. We assumed that no allowances would remain in that bank in 2030. That is, the capped firms together exactly meet the target for emissions by that year.

UCS-NEMS did not allow us to model U.S. links to international cap-and-trade programs to reduce heat-trapping emissions. We were also unable to model any "leakage" of emissions: that is, undercounting of emissions stemming from imports and exports of energy-intensive goods.

the inputs (the "cap emissions trajectory"). However, the cumulative emissions over the modeling period will remain the same for both trajectories, which is the important metric for the climate. (See Chapter 7 and Appendix B online for more information on our results.)

3.4. A Cap-and-Trade Policy Alone Is Not Sufficient

A cap-and-trade program would address the failure of the market to account for harm to the climate. However, it cannot overcome all the barriers to the development and use of technologies and other measures that are essential to creating a true low-carbon economy.

The nation must implement parallel policies alongside a cap-and-trade program, to ensure development and deployment of the full range of energy efficiency and clean energy technologies. These policies—outlined in Chapters 4, 5, and 6—include requiring utilities to generate a higher percentage of their electricity from renewable sources, requiring automakers to increase the fuel economy of their vehicles, stronger energy efficiency standards, incentives for investments in low-carbon technologies, and policies that encourage smart growth, among others.

The results of our analysis provide clear evidence that a comprehensive approach that includes these parallel policies would save households and businesses money by lowering their electricity and gasoline bills, reduce the price of allowances, and help cut heat-trapping emissions.

Finally, a program targeted at reducing such emissions may not, by itself, address other types of local and regional air pollution. We will therefore continue to need strong policies to curb those emissions.

TABLE 3.1. Yearly Caps on U.S. Global Warming Emissions under the Climate 2030 Blueprint

Year	Emissions Cap (million metric tons CO_2 equivalent)
2010	**7,150**
2011	6,501
2012	6,418
2013	6,325
2014	6,221
2015	6,103
2016	5,973
2017	5,830
2018	5,672
2019	5,501
2020	5,317
2021	5,121
2022	4,914
2023	4,699
2024	4,476
2025	4,249
2026	4,021
2027	3,793
2028	3,570
2029	3,353
2030	3,145

The table summarizes year-by-year caps on emissions that were inputs into UCS-NEMS, as key components of the Blueprint cap-and-trade program. The program would begin in 2011.

A program that reduces carbon emissions may not, by itself, address other types of local and regional air pollution that contribute to asthma and other serious respiratory and cardiovascular illnesses in nearly all of the country's major cities. For example, the Philadelphia-Camden-Vineland area (shown here) was ranked the United States' tenth most ozone-polluted metropolitan region in 2006, and global warming is expected to worsen air quality in the region. Strong policies designed to directly curb toxic air pollutants must be additional to any federal program targeting carbon emissions.

BOX 3.2.

How It Works: Cap and Trade

Existing cap-and-trade programs provide important lessons about the need for robust design features. A brief review of real-world experience will illustrate two of these lessons. First, a cap must be tight enough to achieve significant cuts in emissions. Second, the method regulators select for distributing emissions allowances to firms is critical, and auctioning is gaining favor as the preferred approach.

Cap and Trade in Practice

The European Union's Emission Trading Scheme (EU ETS) is the first cap-and-trade program for reducing heat-trapping emissions, and is designed to help European nations meet their commitments to the Kyoto Protocol. This program includes 27 countries and all large industrial facilities, including those that generate electricity, refine petroleum, and produce iron, steel, cement, glass, and paper.

The first phase of the EU ETS—from 2005 to 2007—drew criticism for not achieving substantial cuts in emissions, and for giving firms windfall profits by distributing carbon allowances for free. These criticisms are valid. However, the EU viewed Phase 1 as a trial learning period. The extent to which Phase 2—which runs from 2008 to 2012—helps Europe fulfill its Kyoto commitments will be a better test of the program.

Phase 1 allowed countries to auction up to only 5 percent of allowances—and only Denmark chose to auction that amount. The result was billions of dollars in windfall profits for electricity producers. Phase 2 allows slightly more auctioning, which is expected to occur.

The rules for Phase 3—which extends from 2012 to 2020—were published in December 2008, and unfortunately they are not as ambitious as expected, given the EU's stated commitment to tackling global warming. This phase targets a 20 percent reduction in emissions from 1990 levels by 2020; climate experts had hoped for 30 percent. Even this target is considerably watered down because of the large amount of offsets allowed from outside the capped region. Auctioning of allowances is still not likely to play a major role. This experience reinforces the fact that the United States would be much more likely to win stronger commitments from the EU and elsewhere if it fulfilled its responsibility to lead on climate policy.

The Regional Greenhouse Gas Initiative (RGGI) is a cap-and-trade program that covers a single sector—electricity generation—in 10 northeastern and mid-Atlantic states. The program aims to achieve a 10 percent reduction in emissions from power plants by 2018.

The program's most notable aspect is that states unanimously chose auctioning to distribute the vast majority of emissions allowances. Six of the 10 states will auction nearly 100 percent of their allowances. The auctions of the other four states include fairly small portions of fixed-price sales or direct allocations.

> **"At a time when jobs are being cut all over the country, investments in the clean-energy industry represent just the type of 'jobs program' we need in New Jersey—money-saving, pollution-cutting, and technologically innovative."**
> —Governor Jon Corzine

The program's initial three-year compliance period begins in 2009, but the first multistate auctions occurred on September 25 and December 17, 2008. The first auction, which included allowances from only six states, raised $38.5 million, while the second raised $106.5 million. States and electric utilities will invest the vast majority of those funds in energy efficiency and renewable technologies, with an emphasis on reducing demand for fossil-fuel-based electricity and saving consumers money.

The RGGI auction includes a reserve price, to ensure that CO_2 emissions will always carry a minimum cost, and that the auctions will yield a minimum amount of revenue for these important programs. Some analysts fear that the states may have set the cap too high, because emissions have not grown at the rate expected when the cap was set in 2005. However, there is a possibility that the states could revisit the cap.

Cap and Trade on the Horizon

The Western Climate Initiative (WCI)—which includes seven western states and four Canadian provinces—has established a regional target for reducing heat-trapping emissions of 15 percent below 2005 levels by 2020. WCI's main focus is developing a regional

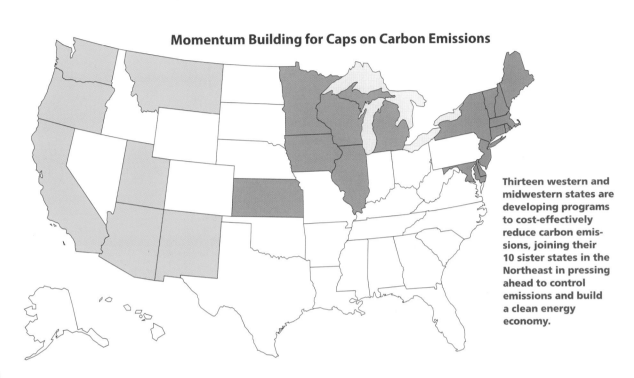

Momentum Building for Caps on Carbon Emissions

Thirteen western and midwestern states are developing programs to cost-effectively reduce carbon emissions, joining their 10 sister states in the Northeast in pressing ahead to control emissions and build a clean energy economy.

cap-and-trade program. The WCI also requires participants to implement California's Clean Car Standard, and recommends other policies and best practices that states and provinces can adopt to achieve regional goals for cutting emissions.

The first phase of WCI development culminated on September 23, 2008, with the release of its Design Recommendations. These sketch out a very broad cap-and-trade program that would cover 85–90 percent of all heat-trapping emissions from participating states and provinces. The only parts of the economy that would remain uncapped are agriculture, forestry, and waste management. However, some sectors, such as transportation fuels, would be brought in at the start of the second compliance period, in 2015.

California is the largest single entity in the WCI, and it has the most detailed action plan of any state in the nation. In 2006 the legislature passed, and Governor Schwarzenegger signed, a law to reduce emissions economywide. The California Air Resources Board has created a blueprint for achieving the required reductions. The plan includes a strong set of sector-specific policies forecast to provide about 80 percent of the needed reductions, as well as a broad cap-and-trade program linking to the WCI. The California and WCI

cap-and-trade programs are scheduled to go into effect in 2012.

Another nascent regional effort is occurring in the Midwest. On November 15, 2007, the governors of Illinois, Iowa, Kansas, Michigan, Minnesota, and Wisconsin, as well as the premier of the Canadian province of Manitoba, signed the Midwestern Regional Greenhouse Gas Reduction Accord. Participants agreed to establish regional targets for reducing global warming emissions, including a long-term target of 60–80 percent below today's levels, and to develop a multisector cap-and-trade system to help meet the targets.

Participants will also establish a system for tracking global warming emissions, and implement other policies to help reduce them. The governors of Indiana, Ohio, and South Dakota joined the agreement as observers. The regional accord for reducing such emissions is the first in the Midwest.

The governors and premier assembled an Advisory Group of more than 40 stakeholders to advise them, and their final recommendations are due in May 2009. As now conceived, the cap would take effect January 1, 2012.

Austin, TX, City Hall

CHAPTER 4

Where We Work, Live, and Play: Technology for Highly Efficient Industry and Buildings

The energy used to power, heat, and cool our homes, businesses, and industries is the single largest contributor to global warming in the United States. Nearly three-quarters of all U.S. energy consumption—and two-thirds of all U.S. carbon emissions—come from those sectors. Fortunately, our industries and buildings are also where some of the most significant and readily available global warming solutions can be found. And no solution is more important to a comprehensive strategy for cutting emissions than energy efficiency.

Energy efficiency technologies allow us to use less energy to get the same—or higher—level of production, service, and comfort. We can still light a room, keep produce fresh, and use a high-speed computer, but we can do it with less energy. Energy efficiency is an appealing strategy because it can yield quick, significant, and sustained energy savings, which typically provide substantial long-term economic returns for consumers and businesses. But technology cannot do it alone. Creating a highly energy-efficient economy also requires policies and programs to help overcome significant, entrenched barriers, and to help businesses and consumers make wise decisions and find ways to eliminate wasteful and unnecessary uses of energy.

Our analysis relied on a supplemental analysis by the American Council for an Energy-Efficient Economy (ACEEE) of the costs and energy savings resulting from policies and programs aimed at spurring the use of energy-efficient technologies in the residential, commercial, and industrial sectors. We used the energy savings resulting from the ACEEE analysis to reduce electricity and fossil fuel use in UCS-NEMS. The model then determined the effects of the cuts in energy use on electricity generation, fossil fuel used to produce electricity, carbon dioxide emissions, energy prices, and energy bills resulting from those policies.[26]

This chapter explores some of the key energy-efficient technologies and innovations that will have the greatest effect in reducing heat-trapping emissions during the coming decades. The chapter then examines the potential for deploying these technologies on a large scale, their associated costs and savings, key challenges and barriers to reaching their full potential, and the suite of policies that the Blueprint supports to help drive their use.

4.1. Energy Efficiency Opportunities in Industry

The industrial sector is an essential component of the U.S. economy, producing millions of different products for consumers each year. That production currently uses a tremendous amount of energy. Industry is responsible for about one-third of all U.S. energy consumption—more than any other sector of the economy—and is also America's second-largest consumer of coal, primarily in the steel, chemicals, and pulp and paper industries. As a result, industry is responsible for more than one-quarter of total U.S. CO_2 emissions, including those from the electricity that industry uses (EIA 2009).

Industry is also a highly diverse sector, with processes, equipment, and energy demands across and within various arenas varying widely (Shipley and Elliot 2006). Petroleum refining, chemicals, and primary metals, for example, account for more than 60 percent

---

26 See Appendix C online for more information on the analysis by ACEEE.

Using an innovative design process, Atlanta-based carpet manufacturer Interface decreased its energy consumed per square yard of product by 45 percent. This achievement is part of a broader vision for sustainability that Interface founder Ray Anderson and his team have parlayed into a global leadership position in the carpet tile industry.

of all energy consumption in the industrial sector. Other industries—such as computers, electronics, appliances, and textiles—are far less energy intensive (EIA 2005). Many of the opportunities for boosting energy efficiency are therefore industry- and site-specific. Achieving our national goals for reducing emissions, then, requires identifying and capitalizing on both industry-wide and site-specific opportunities to deploy energy-efficient technologies and practices.

Numerous studies show an abundance of cost-effective energy efficiency solutions across all industries (Creyts et al. 2007; Nadel, Shipley, and Elliott 2004; IWG 2000). Some of the best opportunities include replacing existing equipment, pursuing innovations in more efficient processes and production technologies, using combined-heat-and-power systems, and relying on recycled petroleum feedstocks.

4.1.1. Equipment Replacement
The electric motor accounts for more than two-thirds of all industrial consumption of electricity (EIA 2008a). Investing in more efficient motors has historically provided significant gains in industrial efficiency—but many opportunities for upgrading today's equipment remain. Improving how companies maintain and coordinate their in-house motor systems can also save energy (Shipley and Elliott 2006). Retrofits to compressed-air systems, heating, ventilating, and air conditioning systems, furnaces, ovens, boilers, and lighting can provide further efficiency gains (Ehrhardt-Martinez and Laitner 2008).

4.1.2. Innovation in Industrial Processes
Some of the best options for boosting energy efficiency involve integrating new technologies into industrial processes. Advanced sensors, wireless networks, and computerized controls optimize energy use while also providing other benefits, such as higher productivity, greater quality assurance, and reduced waste of materials and other inputs (Ondrey 2004). Companies can also reap significant savings by redesigning entire processes to make them more efficient.

4.1.3. Combined-Heat-and-Power Systems
Combined heat and power (CHP) is a well-established but underused technology that entails generating electricity and heat from a single fuel source—dramatically increasing energy efficiency. By recovering and reusing the waste heat from producing electricity, CHP systems can achieve efficiencies of up to 80 percent, compared with about 33 percent for the average fossil-fueled power plant.

Continued advances in CHP and other thermal systems—such as even more effective recovery of waste heat, and the use of such systems for cooling and drying—stand to contribute significant energy savings and cuts in carbon emissions by 2030. Much of the remaining potential lies in industries that have traditionally used CHP, including pulp and paper, chemical, food, primary metals, and petroleum refining. However, industries such as textiles, rubber and plastics, and metal fabrication have considerable untapped potential for using smaller CHP systems (EIA 2008a; EIA 2000).

4.1.4. Recycled Petroleum Feedstocks
Sources of energy not only power industry but also serve as an ingredient—or feedstock—in manufacturing processes. The largest use of petroleum in the manufacturing sector, for example, is as a feedstock in the production of chemicals and plastics. Natural gas, meanwhile, is a key feedstock in the production of fertilizers. Improved techniques and processes that replace virgin petroleum with high-quality recycled or alternative feedstocks are poised to play an important role in reducing carbon emissions.

4.2. Energy Efficiency Opportunities in Residential and Commercial Buildings
The energy used in the buildings where we live, work, shop, meet, and play contributes significantly to our carbon emissions. The residential and commercial sectors account for 21 percent and 18 percent, respectively, of total U.S. energy use as well as CO_2 emissions,

including emissions from electricity used in buildings (EIA 2009). Both sectors use energy primarily to heat and cool spaces, heat water, provide lighting, and run refrigerators and other appliances and electronics (see Figure 4.1). A wealth of readily available solutions for each use could reduce consumption and carbon emissions without sacrificing comfort or quality.

4.2.1. Heating and Cooling

Heating and cooling accounts for nearly half of the average energy consumed in homes—in the form of electricity, gas, and oil—and 43 percent of that used in commercial buildings. Leaks in the average building envelope mean that up to 30 percent of this energy is lost (EERE 2006).

To keep more heat in during winter and more heat out during summer, existing and new structures can be outfitted with better and more appropriate insulation in walls, ceilings, and basements and around ductwork. Highly efficient windows with multiple panes, low-emissivity glass, and insulated frames can also reduce heating and cooling energy use by 20–30 percent (EERE 2006). Radiant barriers—a layer of reflective material in a roof that prevents heat transfer—can also moderate seasonal temperature exchanges in attic spaces, while lighter-colored rooftops can reduce unwanted solar heat gain in warmer climates.

Next to buttoning up a building's envelope, the use of highly efficient equipment can have the biggest impact on reducing carbon emissions from heating and cooling. Owners can easily install ultra-high-efficiency boilers, furnaces, and air conditioners already available

Simple, common-sense decisions often make a significant difference in the long run. Light-colored roofs, like this one at Atlanta's Energy and Environmental Resource Center, reflect sunlight, keeping buildings cooler, reducing demand for air conditioning, lowering electricity use, and saving money.

in new buildings, or in existing structures when equipment wears out. Because most equipment is typically built to last 15 to 25 years, the most efficient models can provide significant long-term energy savings.

Most heating systems use natural gas, oil, or electricity as an energy source, but several existing and

FIGURE 4.1. Residential and Commercial Energy Use

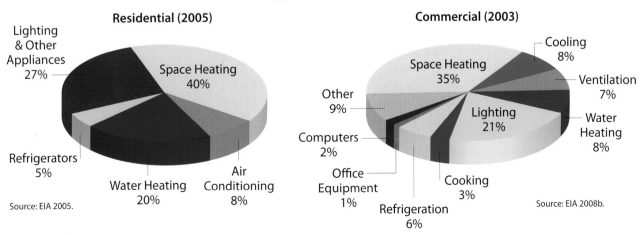

Residential (2005)

Lighting & Other Appliances 27%
Space Heating 40%
Refrigerators 5%
Water Heating 20%
Air Conditioning 8%

Source: EIA 2005.

Commercial (2003)

Space Heating 35%
Cooling 8%
Ventilation 7%
Water Heating 8%
Lighting 21%
Other 9%
Computers 2%
Office Equipment 1%
Refrigeration 6%
Cooking 3%

Source: EIA 2008b.

Space heating and cooling account for the largest portion of home and business energy budgets. Lighting, water heating, and refrigeration are also substantial energy consumers in buildings. Fortunately, there are significant opportunities for energy and cost savings through efficiency.

emerging sources offer greater efficiency. For example, geothermal (ground-source) heat pumps use the constant temperature below ground to provide heating or cooling with much less energy. Air-source heat pumps, which use the difference between outdoor and indoor air temperatures for cooling and heating, are also effective in more moderate climates. Micro-combined-heat-and-power systems are also an emerging option that can allow commercial buildings and homes to get the most out of their fuel use. Similar to larger systems, micro-CHP meets heating and even cooling needs with the excess heat from on-site electricity generators powered, for instance, by natural gas.

Several other solutions from simple to high-tech can also help save energy and cut carbon emissions from heating and cooling. Ceiling fans can significantly reduce the need for air conditioning, and programmable thermostats (which can even be controlled remotely) can reduce energy use by 5–15 percent. Passive solar designs can minimize energy use and increase the comfort of new buildings by considering the sun's location

Energy efficiency in buildings generates many types of jobs—for contractors, plumbers, and electricians who renovate existing buildings as well as engineers and architects who design new ones. Some architects specialize in passive solar design that decreases a building's lighting and heating needs.

at various times of year. For example, large south-facing windows with good overhangs can let winter sun in and keep summer sun out. Well-placed trees can also help shade buildings from the high summer sun and protect them from winter winds.

4.2.2. Water Heating

Water heating offers strong opportunities for cutting carbon emissions, as it accounts for about 20 percent of energy used in residential buildings, and 8 percent of energy used in commercial buildings (EIA 2008b; EIA 2005). High-efficiency water heaters that are available today use 10–50 percent less energy than standard models, and new advances are expected to offer further gains (EPA 2008b).

On-demand or "tankless" water heaters, which heat water only when it is needed, reduce energy consumption 10–15 percent by avoiding "standby" losses (Amann, Wilson, and Ackerly 2007). Innovations in gas-condensing water heaters—which capture and use warm combustion gases to heat water further, before releasing the gases to the outdoors—can reduce the amount of energy used to heat water by as much as 30 percent (EPA 2008b).

Fuel choice is also important for curbing carbon emissions. Natural-gas-fired water heaters are far more efficient than those powered by oil or electricity, if we account for the inefficiencies that occur producing the electricity. However, solar water heaters offer the greatest cuts in carbon emissions. Innovations in the design of such systems have improved their efficiency, significantly reduced their cost, and allowed their use in most climates.

4.2.3. Lighting

Lighting accounts for about 10 percent of an average home's energy use, and more than 20 percent of the energy used in the commercial sector (EIA 2008b; Amann, Wilson, and Ackerly 2007). Large-scale changes to the lighting industry now under way will deliver significant cuts in energy use and carbon emissions.

A provision in the Energy Independence and Security Act of 2007 (EISA) requires lightbulbs to be 30 percent more energy efficient starting in 2012, with further reductions mandated by 2020. These new standards will effectively phase out traditional incandescent bulbs.[27] Their replacements will be compact fluorescent

---

27   Our Reference case included the lighting efficiency standard and other provisions in EISA.

lightbulbs (CFLs), light-emitting diodes (LEDs), and advanced incandescent lamps that use halogen capsules with infrared reflective coatings now in development. EISA's provision for efficient lightbulbs is projected to reduce annual U.S. carbon emissions 28.5 million metric tons by 2030 (ACEEE 2007).

Gas discharge lamps—such as metal halide and sodium vapor—which pass electricity through gases to produce light, are two to three times more efficient than CFLs, and thus save even more energy. These lamps are typically used in office buildings and retail outlets because of their large size. However, technological advances are broadening their application to smaller-scale residential uses.

Of course, lighting uses the least amount of energy when it is turned off. Building designs that maximize natural light from the sun (known as daylighting) through the use of windows, skylights, and glass partitions can significantly reduce energy use in both residential and commercial settings. Sensors that adjust lamp output based on ambient lighting conditions, and automatically turn off lights in empty rooms, can also help cut global warming emissions.

4.2.4. Appliances and Electronics

Large appliances such as refrigerators, washing machines, and dishwashers account for about 20 percent of household energy use. Electronics comprise a smaller but growing share of electricity demand—primarily because of the rapid growth of larger television screens, faster computers, video games, and handheld devices such as cell phones and MP3 players.[28] Manufacturers have made great strides in enabling many of these products to run on less power. For example, innovations in motors, compressors, and heat exchangers, as well as better insulation, have made today's refrigerators three times more efficient than their 1970s counterparts (Nadel et al. 2006).

High-efficiency models of most appliances and electronics are available today. The models highlighted by the U.S. Environmental Protection Agency's Energy Star program typically offer energy savings of 20 percent or more. Electronics manufacturers are also continuing to research and design equipment, appliances, and gadgets that are more energy efficient. These rely on ever-smaller microprocessors for computers, organic

Standards designed to increase the energy efficiency of home appliances and electronics help consumers save money on electricity bills by reducing energy demand. For example, America's 275 million televisions consume more than 50 billion kilowatt-hours of electricity each year—equivalent to the output of more than 10 coal-fired power plants. Efficient Energy Star televisions use 30 percent less power.

LEDs (which use a thin film made from organic compounds) for lighting large-screen TVs, and microhydrogen fuel cells to replace lithium-ion batteries.

4.2.5. On-Site Generation of Clean Electricity

Homes and businesses can also reduce carbon emissions by using clean and renewable resources to generate electricity on-site. Solar electric systems (known as photovoltaics, or PV) are an option for any building with good access to the sun. Advances in technology are also opening up new opportunities to integrate PV into buildings directly—in place of shingles, façades, skylights, or windows. Small-scale wind systems may also be an effective option for generating carbon-free electricity on-site, particularly in rural areas.[29]

4.3. Potential for Greater Efficiency

Potential

Energy efficiency has already been working hard and providing significant dividends to the U.S. economy for nearly four decades. A recent study found that energy-efficient technologies and practices have actually met *three-quarters* of all new demand for energy services since 1970 (see Figure 4.2). Over that same period, the energy intensity of the U.S. economy—that

---

28 The appreciable amount of energy used by many household electronics when not in operation is another opportunity. These standby energy losses—also known as "vampire" or "phantom" losses—add up to some 65 billion kilowatt-hours of electricity per year, or about 5 percent of residential electricity use. See *www.ucsusa.org/publications/greentips/energy-vampires.html.*

29 Chapter 5 and Appendix D (available online) describe renewable energy technologies in greater detail.

FIGURE 4.2. Efficiency Helps Meet U.S. Energy Demand

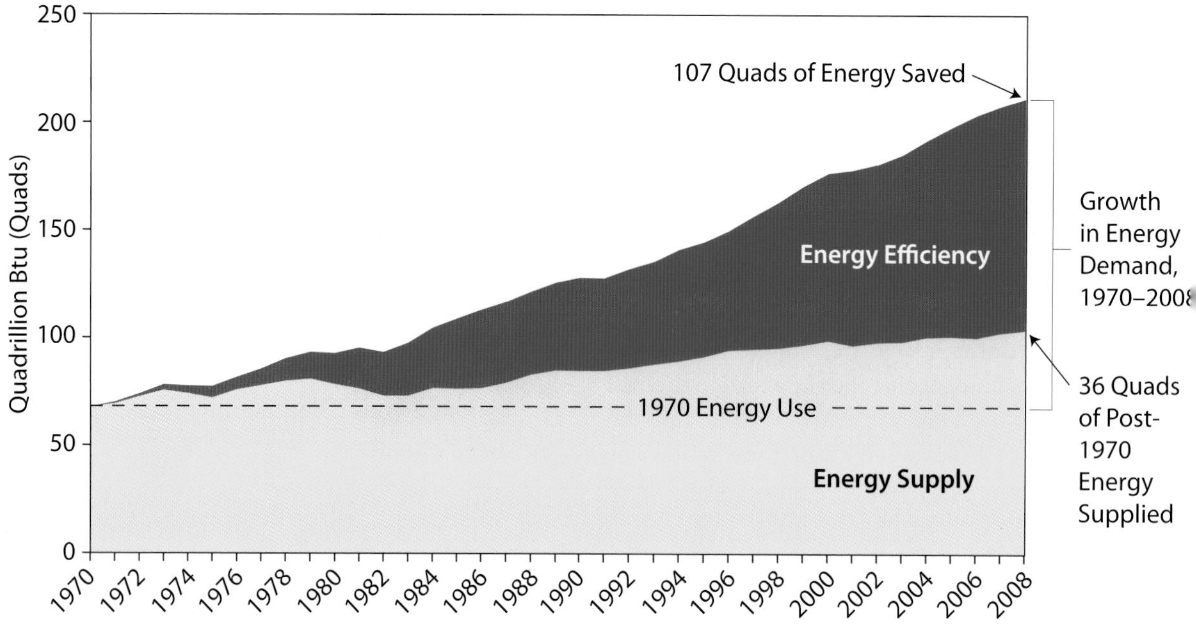

Source: Ehrhardt-Martinez and Laitner 2008.

Over the past four decades, U.S. energy needs have more than tripled. Energy-efficient technologies and practices have been able to meet three-quarters of this demand, sharply reducing the amount of conventional energy resources needed to meet remaining demand. Further advances in energy efficiency have the potential to make even greater cuts in energy use across all economic sectors and within every region of the country.

is, energy consumption per dollar of economic input—has fallen by more than half, largely because of improved efficiency (Ehrhardt-Martinez and Laitner 2008). Yet despite these important successes, energy efficiency is an underused resource in the United States. A massive reservoir of potential energy efficiency remains untapped, ready to contribute to the challenge of reducing our carbon emissions.

Research into the potential of energy efficiency typically considers only measures that are or may become

cost-effective, rather than the full—or "technical"—potential. A recent meta-analysis of 11 studies at the state and national level found that the technical potential for reducing energy use from efficiency measures is 18–36 percent for electricity, and 38–47 percent for natural gas (see Table 4.1) (Nadel, Shipley, and Elliot 2004).

The greatest potential for reducing the use of electricity through energy efficiency lies in the commercial and residential sectors. For natural gas, the potential for energy efficiency is greatest in the residential sector, specifically in space and water heating.

The nation also has a wealth of untapped potential for using new combined-heat-and-power systems to boost energy efficiency. The industrial sector has installed about 26,000 megawatts of CHP capacity, which now supply about 7.5 percent of all U.S. electricity use. This capacity is dominated by large systems—those that produce more than 20 megawatts—in the pulp and paper, chemical, food, primary metals, and petroleum refining industries (EIA 2008a).

The total technical potential of CHP at industrial facilities today is estimated at 132,000 megawatts (EIA 2000). The commercial sector—including hospitals, schools, universities, hotels, and large office buildings—

TABLE 4.1. Energy Efficiency Potential

(percent reduction in energy use)

Sector	Natural Gas	Electricity
Residential	46–69%	22–40%
Commercial	16–29%	17–46%
Industrial	NA	18–35%
Total, All Sectors	**38–47%**	**18–36%**

Source: Nadel, Shipley, and Elliot 2004.

Note: These reductions represent technical potential. Real-world barriers may prevent these sectors from reaching their full potential.

also has tremendous opportunities to deploy CHP systems. The total technical potential of CHP in this sector is some 77,000 megawatts (EIA 2000).

4.4. Costs of Improving Energy Efficiency

Understanding the technical potential of energy efficiency can offer an upper bound on the role it can play in helping to reduce global warming emissions. However, the solutions that prove the most economical are the most likely to be developed. Technologies and practices that improve energy efficiency tend to be more cost-effective than other global warming solutions—which is why efficiency must be the cornerstone of any comprehensive strategy for cutting carbon emissions.

Over time, reductions in energy use more than offset the initial costs of most efficiency solutions—so they often provide significant long-term economic benefits. By reinvesting some of the money saved on energy bills, the nation can afford to invest in other critical global warming solutions that may be more expensive.

For example, a 2007 analysis by McKinsey & Company found that measures and technologies that

FIGURE 4.3. The Energy Savings and Costs of Efficiency Programs

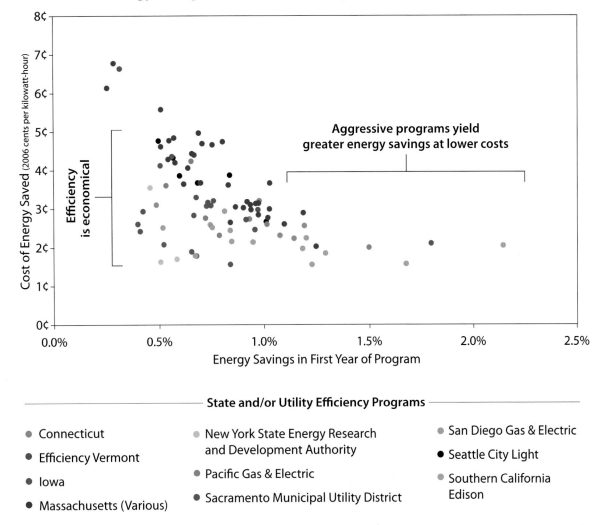

State and/or Utility Efficiency Programs

- Connecticut
- Efficiency Vermont
- Iowa
- Massachusetts (Various)
- New York State Energy Research and Development Authority
- Pacific Gas & Electric
- Sacramento Municipal Utility District
- San Diego Gas & Electric
- Seattle City Light
- Southern California Edison

Source: Adapted from Hurley et al. 2008.

A review of utility- and state-level efficiency programs found that the cost of implementing energy efficiency measures ranged from about 1.5 cents to nearly 7 cents per kilowatt-hour (¢/kWh) saved, with a median of 3.0 ¢/kWh. This is lower than the average U.S. retail price for electricity (about 9.1 ¢/kWh). The review also found that implementation costs are cheaper when a program enables greater efficiency gains. This suggests that an aggressive, comprehensive plan to boost energy efficiency nationwide—as recommended in the Blueprint—is the most cost-effective approach and would provide the greatest benefits for consumers.

BOX 4.1.

SUCCESS STORY

The Two-Fer: How Midwesterners Are Saving Money while Cutting Carbon Emissions

In every region of the country, people are seeing the advantages of improving energy efficiency in residential buildings. Single-family homes, apartment buildings, and even entire neighborhoods can be built new or renovated to boost energy efficiency—saving families money while reducing heat-trapping emissions.

Cleveland may, at first blush, seem an unlikely place to find green homes. The post-industrial city suffers from severe winters, residential flight, and industrial decline. By adding Cleveland to the emerging midwestern "Green Belt"—a reference to the region's moniker as the Rust Belt—the city's residents, businesses, and government see an opportunity to attract new industries and reverse population decline.

Artfully mixed with historic housing, the new energy-efficient homes in Cleveland's EcoVillage add to the diversity of Detroit-Shoreway—a neighborhood of mostly renting families, with a few young professionals and "empty nesters." Believing that a stable neighborhood is a socioeconomically mixed one (Hansen 2008), EcoVillage designers worked with the community to integrate these new homes into the fabric of the neighborhood (Metcalf 2008).

The 20 new village townhouses and two single-family homes sold for close to median market prices. Five "green" cottages will be made available to residents making less than 80 percent of Cleveland's median income (Dawson 2008). All the homes are equipped with energy-efficient appliances, double-pane windows, extra insulation, and high-performance heating, cooling, and air conditioning systems, to reduce energy use and utility bills.

Some units take advantage of passive solar heating through south-facing windows, and were built with framing that leaves space for more insulation (Metcalf 2008). Four of the townhouses also have photovoltaic panels on their garages, supplying a substantial percentage of each home's electricity needs. Reports Mandy Metcalf, former EcoVillage project director, "A couple of the homeowners that have the panels were getting negative energy bills, actually getting credits on their energy bills" (Metcalf 2008).

Thanks to these simple construction techniques and the use of energy-efficient products—which are available around the country for competitive prices—heating bills for residents of EcoVillage are drastically lower than those for residents of standard housing. For example, heating costs for one of the three-bedroom green cottages are projected to be only $432 per year—less than half the amount a typical midwestern household expected to spend during the 2008–2009 winter (Cuyahoga Land Trust 2008; EIA 2008c).[30]

Ohio's EcoVillage cottages (left) and Minnesota's Viking Terrace apartments (right) are good examples of how energy efficiency and smart building design can save money and reduce carbon emissions all around the country.

The renovation of Viking Terrace, an income-based rental complex in rural Minnesota, is another green housing success story. With funding from the city and federal governments, nonprofit organizations, and low-income housing tax credits, the Southwest Minnesota Housing Partnership renovated 60 dilapidated apartments into energy-efficient, clean, safe, and affordable housing. The apartments are now equipped with Energy Star appliances and windows, improved insulation, water-conserving appliances and fixtures, a new ventilation system, and a metal gable roof (Minnesota Green Communities n.d.).[31] Renovators also installed a high-efficiency geothermal heating and cooling system—the project manager's proudest, and largest, investment (Lopez 2008). The partnership expects this system to pay for itself through energy savings in just a decade, and tenants say they love it (Lobel 2007).

With these installations, the partnership expects to cut household energy and water use by 40 percent (Buntjer 2007)—a significant decrease in the harsh Minnesota climate. Today all 60 apartments are happily occupied, and 15 families are on the waiting list. Four of the apartments are affordable to families earning 30 percent of the area's median income, while 47 are affordable to families earning 50 percent of the median (Minnesota Green Communities n.d.)—a strong testament to the desirability and economic benefits of green renovations.

provide positive economic returns could provide nearly 40 percent of the cuts in carbon emissions required by its mid-range case. Of these cost-effective solutions, nearly 60 percent stem directly from energy efficiency gains in industry and buildings. McKinsey's mid-range case projects that making buildings and industry more efficient could reduce U.S. demand for electricity 24 percent by 2030. That, in turn, could provide one-third of the needed reductions in CO_2 emissions, at an average weighted net savings of $42 per ton of CO_2 equivalent (in 2005 dollars) (Creyts et al. 2007).

Our analysis of policies to promote energy efficiency shows that they can reduce total U.S. energy consumption 29 percent (12 quadrillion Btu, or 12 quads) by 2030—or an average of 1.3 percent per year. We assumed that the annual costs of those policies would reach $7.5 billion in 2020, and rise to $13.4 billion in 2030. Those costs include expenditures related to developing and administering programs, research and development, and incentives to encourage households and businesses to boost energy efficiency. Those expenditures, in turn, stimulate $64.3 billion in new spending for more energy-efficient technologies and measures in 2020, and $113.6 billion in 2030. (See Table 4.3 for a breakdown of policy and investment costs.) The levelized cost of these investments in energy efficiency would be about $12.62 per million Btu.[32]

Other recent studies also suggest that energy efficiency could cost-effectively reduce U.S. energy use 25–30 percent over the next 20 to 25 years, or 1–1.5 percent per year (Ehrhardt-Martinez and Laitner 2008; ASES 2007; Nadel, Shipley, and Elliott 2004; IWG 2000).

Leading state energy efficiency programs have already achieved such annual cuts in energy use. For example, energy efficiency programs in Vermont reduced electricity use by more than 1.7 percent in 2007, and have averaged cuts of more than 1.1 percent since 2003 (Efficiency Vermont 2007). California has also seen aggressive reductions: per capita electricity use has remained constant in that state since the mid-1970s, while rising nearly 50 percent in the country as a whole (CEC 2007).[33] During California's energy crisis in 2001, about one-third of the 6 percent drop in electricity use came from

---

30 The Energy Information Administration projected that the average midwestern household would spend $1,056–$1,175 on heat during the winter of 2008–2009. That range reflects the different prices of heating fuels. The cost of heating with electricity was expected to be $1,056, while the cost of heating with propane was projected to be $1,941. The cost of heating with natural gas and oil fell within this range (EIA 2008c).

31 Pumping, distributing, treating, and heating water takes energy. Running a standard hot water faucet for five minutes requires about as much energy as keeping a 60-watt lightbulb lit for 14 hours (City of Chicago 2008), and water heating alone accounts for 13–17 percent of a typical household's utility bill (EERE 2009a).

By reflecting light and heat back into the air rather than absorbing and transferring it to the house below, as traditional black roofs do, metal roofs can substantially reduce the energy required to cool houses. According to the Energy Star program, qualified reflective roofing can lower surface temperatures by up to 100°F, and reduce peak cooling demand by 10–15 percent (Energy Star 2009).

---

32 The levelized cost is the annualized cost of the total efficiency investment divided by the total savings.

33 While California's steady per capita electricity use likely stems from a range of factors, its early energy efficiency policies were a major factor in enabling the state to meet growth in energy demand with greater efficiency (Sudarshan and Sweeney 2008).

investments in energy-efficient technologies (Global Energy Partners 2003).[34]

Reducing energy use a minimum of 1 percent per year is consistent with key commitments by leading states. California, Connecticut, and Michigan all require annual savings in electricity use of 1 percent. Other states and regions have adopted even higher requirements, including Minnesota (1.5 percent), Maryland (~2 percent), Illinois (2 percent starting in 2015), Ohio (2 percent starting in 2019), and the Midwestern Greenhouse Gas Reduction Accord (2 percent).[35]

A recent review of 14 utilities, groups of utilities, and state efficiency programs found that the cost of measures for making electricity use more efficient ranged from about 1.5 cents to nearly seven cents per kilowatt-hour saved, with a median of three cents per kilowatt-hour (Hurley et al. 2008). That analysis also uncovered a correlation between the cost of reducing energy use and the size of the program. That is, energy savings are cheaper when a program itself achieves greater efficiency (Hurley et al. 2008).

Energy use in existing buildings represents a significant portion of residential and commercial electricity demand. Because most buildings standing today will still exist in 2030, energy-saving improvements such as additional insulation or replacement windows will be necessary to reduce the carbon emissions associated with these buildings.

This finding suggests that an aggressive, comprehensive plan to boost energy efficiency nationwide could benefit from economies of scale as well as more effective coordination (Hurley et al. 2008). Indeed, while cuts in energy use from some mature efficiency technologies might decline with more widespread use, our analysis assumes that any diminishing returns would be more than offset by economies of scale and the introduction and growth of newer technologies.

4.5. Key Challenges for Improving Energy Efficiency

Challenges

Despite clear economic and environmental advantages, energy efficiency still faces many market, financial, and regulatory barriers to achieving its full potential. One of the steepest market barriers is the "split incentive" (Prindle et al. 2007). That is, builders of new homes and businesses have a strong motivation to keep construction costs low, and little incentive to optimize a building's efficiency, as buyers will be the ones paying for energy use. Landlords are similarly less interested in investing in energy efficiency when tenants reap most of the benefits (Ehrhardt-Martinez and Laitner 2008).

Lack of information among energy consumers is another common challenge. They may not be aware of, or simply underestimate, the impact of the efficiency of their purchases—whether a handheld gadget, major appliance, or even a house—on energy use. Such information is often not readily available, and consumers may not have the time, ability, or inclination to do the required research. And at companies and large institutions, maintenance staff or other employees who lack complete information—or who place a higher priority on keeping capital costs low than on overall costs—often make purchasing decisions (Nadel et al. 2006).

Higher-efficiency products also typically have higher up-front costs than their counterparts. Homeowners and businesses may lack the capital or financing to make larger initial investments. And publicly traded corporations focused on showing profits to shareholders are often unwilling to make investments in energy efficiency that do not produce significant near-term returns.

Particular technologies or approaches to energy efficiency face additional barriers. Despite the clear

---

34   The remainder resulted from aggressive conservation measures.

35   The Midwestern Greenhouse Gas Reduction Accord is a regional agreement by governors of six states (Illinois, Iowa, Kansas, Michigan, Minnesota, and Wisconsin) and the premier of Manitoba to reduce emissions to combat climate change. For more information, see Box 3.2.

TABLE 4.2. Energy Savings in Buildings and Industry from Blueprint Policies

Blueprint Policies	Electricity Savings (billion kilowatt-hours)		Total Energy Savings[a] (quadrillion Btu)	
	2020	2030	2020	2030
Appliance and Equipment Standards	104	193	1.01	1.75
Energy Efficiency Resource Standard	390	652	2.17	3.68
Energy Efficiency Codes for Buildings	131	223	0.76	1.25
Advanced-Buildings Program	69	168	0.46	1.06
R&D on Energy Efficiency	18	200	0.17	1.76
Combined-Heat-and-Power Systems[b]	264	453	0.34	0.58
Energy-Efficient Industrial Processes	51	100	0.89	1.73
Enhanced Rural Energy Efficiency	3	3	0.01	0.01
Use of Recycled Petroleum Feedstocks	—	—	0.16	0.26
Total	**1,030**	**1,992**	**5.97**	**12.08**

The suite of Blueprint efficiency and combined-heat-and-power policies deliver strong energy savings by 2020, and by 2030, the efficiency gains double in size.

Notes:

a Total energy savings include reductions in the use of electricity as well as natural gas, home heating oil, and other sources of energy.

b Total energy savings for combined heat and power include more widespread use of natural gas in the commercial and industrial sectors, equal to 0.56 quadrillion Btu.

economic advantages of CHP, for example, significant regulatory and market barriers that discourage power producers other than utilities are preventing it from achieving its full potential. For example, developers of CHP projects seeking to connect with the electricity grid often face discriminatory pricing and technical hurdles by uncooperative utilities (see Brooks, Elswick, and Elliott 2006). High-quality recycled materials that could replace petroleum feedstocks in industry also face market barriers, such as lack of knowledge among manufacturers of how to process those resources.

Cutting carbon emissions swiftly and deeply, meanwhile, will require making existing buildings more energy efficient. New technologies and advanced building designs are usually easier to introduce into new construction. Yet more than 113 million single-family, multi-family, and mobile homes already exist, and commercial buildings have more than 75 billion square feet of floor space (EIA 2009). The vast majority of these buildings will still be in use in 2030, and most will still be standing even in 2050. The nation will need to mount a concerted and coordinated effort—supported by effective public policies—to improve the energy efficiency of these structures.

4.6. Key Policies for Improving Energy Efficiency

Policies

As part of its analysis, the American Council for an Energy-Efficient Economy evaluated the costs and energy reductions of a suite of policies designed to remove key obstacles to maximizing the impact of energy efficiency (see Table 4.2). These policies build on the most effective approaches by leading states and the federal government.

4.6.1. Energy Efficiency Standards for Appliances and Equipment

Appliance and equipment standards save energy by requiring that various new products achieve minimum levels of efficiency by a certain date. As higher-efficiency products gradually enter the market, they replace older, less-efficient models while still offering consumers a full range of options. Such standards help overcome market barriers to more efficient products, such as lack of awareness among consumers, split incentives between developers and buyers (and landlords and tenants) and limited availability of such products.

Efficiency standards have been one of the federal government's most successful strategies for reducing

The new "whole-building" approach to architecture attempts to incorporate energy efficiency and passive solar technologies while creating an attractive, open aesthetic. One impressive example in Michigan, the Grand Rapids Art Museum (shown here both inside and out), meets the gold standard of sustainability criteria established by the U.S. Green Building Council and was named one of *Newsweek*'s Six Most Important Buildings of 2007.

energy consumption in homes and businesses since their inception in 1987. For example, the annual amount of energy saved primarily due to efficiency standards for appliances and equipment reached 1.2 quadrillion Btu (1.3 percent of total energy use) in 2000. By 2020, annual energy savings from today's efficiency standards are projected to grow to 4.9 quads (4.0 percent)—equivalent to the total energy used by some 27 million homes (Nadel et al. 2006).

The Blueprint assumes that the federal government establishes new or upgraded efficiency standards for 15 types of appliances and equipment—including incandescent lamps, electric motors, refrigerators, and clothes washers—over the next several years.

4.6.2. Energy Efficiency Resource Standard

The energy efficiency resource standard (EERS) is emerging as an effective way to promote investment in energy-efficient technologies. Similar to a renewable electricity standard, an EERS is a market-based policy that requires utilities to meet specific annual targets for reducing the use of electricity and natural gas (Nadel 2006). Besides spurring significant cuts in the use of both electricity and natural gas, an EERS can reduce excess demands on the capacity of the grid used to transmit electricity. Some 18 states as well as countries

such as France, Italy, and the United Kingdom have adopted such a standard.

The Blueprint assumes that the federal government sets an EERS that applies to the use of both electricity and natural gas. The electricity target would reduce demand for power by 0.25–1 percent each year, to achieve a total reduction of 10 percent by 2020 and 20 percent by 2030. The natural gas target would eventually reach 0.5 percent annually, reducing use of that energy source a total of 5 percent by 2020 and 10 percent by 2030.[36] Those targets are consistent with standards in leading states such as Minnesota and Illinois, which sometimes set even stricter targets (Nadel 2007).

4.6.3. Energy Efficiency Codes for Buildings

Energy codes for buildings require that all new residential and commercial construction meets minimum criteria for energy efficiency. Adopting more stringent energy codes over time ensures that builders deploy the most cost-effective technologies and best practices in all new construction.

The Blueprint assumes that efficiency codes reduce energy use 15 percent in new residential and commercial construction through 2020, and 20 percent from 2020 to 2030. Those cuts in energy use modestly improve on today's building codes, and are well within

---

36 The EERS does not include any contributions from combined-heat-and-power systems or recycled petroleum feedstocks. This chapter addresses those contributions separately.

the goals recently established by the American Society of Heating, Refrigerating, and Air-Conditioning Engineers (ASHRAE), the American Institute of Architects (AIA), and DOE.

4.6.4. Advanced-Buildings Program

New homes and businesses can save even more energy beyond the cuts prompted by enhanced building codes, if architects design new structures directly for energy efficiency. An advanced-buildings program combines training and technical assistance on new design and construction techniques for architects, engineers, and builders with educational outreach to purchasers on the benefits of energy efficiency. National efforts such as the Environmental Protection Agency's Energy Star program, the U.S. Green Building Council's Leadership in Energy and Environmental Design (LEED) program, and the New Building Institute's Core Performance program encourage builders to incorporate sustainable practices into their construction and help educate consumers.

The Blueprint assumes that a targeted advanced-buildings program gradually ramps up to achieve a 15 percent reduction in energy use by new residential and commercial buildings by 2023, with savings continuing at that level through 2030. This potential is consistent with those considered in other analyses (Elliott et al. 2007a; Sachs et al. 2004).

4.6.5. R&D on Energy Efficiency

Existing knowledge of energy efficiency can lead us far down the path to critical cuts in carbon emissions. However, the scale of the global warming crisis requires us to develop new technologies and practices over the coming decades. Investment in research and development is therefore essential to identifying and commercializing these approaches.

Federal R&D programs have a long history of advancing the performance and lowering the cost of emerging energy-efficient technologies. These programs are also a sound investment of taxpayer dollars, given that the lifetime economic benefits of such technologies typically far exceed their initial cost.[37]

The Blueprint bases cuts in energy use stemming from federal R&D programs on a study of potential reductions in Florida by ACEEE (Elliott et al. 2007b). We scaled up those savings to the national level, and assumed that a concerted national effort could double

them. As a result of that investment, U.S. energy use falls 4.4 percent by 2030—accounting for about 15 percent of all reductions in energy use from greater efficiency, including CHP.

We also assumed that the nation would need to spend $80 million on R&D (in 2005 dollars over a five-year period) to develop a technology that eventually saves 1 million Btu of energy when it first enters the market, based on estimates from a 1997 report by the President's Committee of Advisors on Science and Technology (PCAST 1997). As a result, the Blueprint projects that a federal R&D program would cost nearly $1.8 billion annually in 2020, and more than $4.6 billion annually in 2030. This funding spurs $2.0 billion in private-sector investments in 2020, growing to $18.5 billion in 2030.

4.6.6. Combined-Heat-and-Power Systems

The nation will have to take several steps to reduce the barriers to widespread adoption of CHP. These include establishing:

• Consistent national standards for permitting and connecting CHP systems to the local power grid.

Combined-heat-and-power (CHP) systems are an energy-saving option for every region of the country. In Texas, CHP accounted for more than 21 percent of electric power generation in 2005, and more than 1,500 miles away, the small community of Epping, NH, installed the micro-CHP system shown here in its 125-year-old town hall. This system, integrated with an array of solar panels, has reduced the building's electric bill by 50 percent, its heating costs by 50 to 60 percent, and its carbon emissions by 60 tons per year.

---

37 See, for example, PCAST 1997.

BOX 4.2.

SUCCESS STORY
Three Companies Find Efficiency a Profitable Business Strategy

Regardless of size, location, or product, all companies agree: reducing global warming emissions must be a profitable business strategy. Here is how three companies accomplished that task.

DuPont

Inspired by scientific consensus on the urgency and magnitude of the threat from global warming, chemical manufacturing company DuPont cut its worldwide heat-trapping emissions 72 percent below 1990 levels in just 10 years (Hoffman 2006). The company achieved those drastic reductions first by capturing and destroying its most abundant global warming emissions (DuPont 2008).

The company then turned its attention to making its industrial processes and instrumentation more efficient, and to installing combined-heat-and-power systems (CHP) at a number of sites (Hoffman 2006). These energy-saving techniques paid off: DuPont's energy use fell 7 percent from 1990 to 2006, even while production expanded 30 percent, saving the company $2 billion (Hoffman 2006).

SC Johnson and Son

As a charter member of the EPA's Climate Leader's Initiative, SC Johnson and Son set an initial goal of reducing its domestic global warming emissions by 8 percent. Far surpassing that goal, the company achieved a 17 percent reduction (EPA 2009), and has committed to an additional 8 percent reduction by 2011 (SC Johnson & Son 2008).

The company credits its success to changes in the way it obtains its energy. Starting in Racine, WI, with its largest manufacturing facility—and largest carbon emitter—the company now uses landfill methane and natural gas to power a CHP plant that provides all of the facility's electricity, and more than half of the steam needed for its processes (EPA 2009). Saving the company millions of dollars annually on energy bills, the CHP plant will pay for itself in less than seven years (EPA 2009). The plant has also reduced the facility's

global warming emissions by 52,000 tons per year (CSR 2007).[38]

Harbec Plastics

Near the shores of Lake Ontario in upstate New York, Harbec Plastics, a small local company, is using a similar business strategy to achieve the same success. Facing rising energy costs and frequent power outages, president and CEO Bob Bechtold decided to invest in new systems that would reduce his company's dependence on an unreliable electricity grid while cutting carbon emissions.

Bechtold first replaced the equipment at the core of his business with newer, more efficient machines. To provide reliable power for this equipment, Bechtold next installed a CHP system that more than handles the plant's electricity demand, and supplies heat and air conditioning at no extra cost (Bechtold 2008a). Both the energy-efficient machines and the CHP system required an up-front investment that the company recouped in two to three years through substantially lower energy bills (Bechtold 2008a).

Finally, Bechtold erected a wind turbine on-site to harness the steady wind blowing off the lake. Producing 10 percent of the plant's total electricity needs, the turbine saves the company $40,000 a year, and allows Bechtold to forecast a substantial portion of his energy bill far into the future (Bechtold 2008a).

These efforts have reduced Harbec's global warming emissions by more than 3,077 tons per year, and put the company on track to be carbon-neutral by 2016 (Bechtold 2008b). The cuts in energy use have also improved the company's bottom line: Harbec Plastics has exceeded its profit projections for the past three years despite failing to meet its sales projections (Bechtold 2008b).

These success stories show that up-front investments in energy-saving and energy-producing technologies not only provide significant cost benefits but also reduce heat-trapping emissions. Harbec Plastics, SC Johnson and Son, and DuPont are but three examples

---

38   This is equivalent to taking 7,700 cars off the road, calculated using an average of 6.75 tons of CO_2 emitted per car per year.

Bob Bechtold's use of microturbines within a combined-heat-and-power system is one of several energy innovations helping Harbec Plastics run efficiently and profitably.

of the many companies that have found cutting such emissions compatible with a sound and profitable business strategy.

Regardless of size, location, or product, all companies agree: reducing global warming emissions must be a profitable business strategy.

- Equitable interconnection fees, and tariffs for stand-by, supplemental, and buy-back power, to help over-come discriminatory pricing practices.
- Uniform tax treatment to level the playing field for all CHP systems regardless of their size or use, and to help reduce their initial capital costs.

The Blueprint also includes annual spending on federal and state CHP programs, such as the successful DOE/EPA CHP Regional Application Centers, which spur the use of CHP through education, coordination, and direct project support, such as site assessments and feasibility studies (Brooks, Elswick, and Elliott 2006). Under the Blueprint, the annual, amortized cost of such programs reaches $48 million in 2020, and $59 million in 2030.

The Blueprint assumes that these policies and investments produce 88,000 megawatts of new CHP capacity by 2030—or an average of 4,000 mega-watts each year—representing nearly half of that tech-nology's technical potential. This rate is consistent with increases this decade in states with effective CHP policies, such as Texas. In that state, CHP accounted for more than 21 percent of electric power generation in 2005—a 29 percent increase over 1999 levels (Elliott et al. 2007a).

4.6.7. Energy-Efficient Industrial Processes

Every aspect of the industrial sector has significant po-tential for low-cost improvements in energy efficiency. The key is to optimize the efficiency of the processes used in each industry and at each site (Shipley and Elliott 2006).

Programs that help facilities identify such opportu-nities and develop strategies for implementing them—such as the DOE's Industrial Assessment Centers and its Save Energy Now program—can enable industry to fulfill this potential. The Blueprint assumes that these and similar efforts will expand, and that local programs will support plant-level efforts.

These programs lead to a 10 percent reduction in the amount of fuel used in industry (not otherwise af-fected by the energy efficiency resource standard or CHP policies) by 2030. This target is consistent with the cost-effective cuts identified by the DOE, after evaluating more than 13,000 in-plant assessments con-ducted since 1980 (Shipley and Elliott 2006).

4.6.8. Enhanced Rural Energy Efficiency

Robust programs to improve the efficiency of energy use in agriculture emerged in the 1970s, in response to rising energy costs on this energy-intensive sector of

TABLE 4.3. Key Policies for Improving the Energy Efficiency of Industry and Buildings

	Total Savings in 2030 (in End-Use Quads)	Total Cost in 2030 (in Billions of 2006 Dollars)	
		Program	Investment
Appliance and equipment standards: The federal government upgrades energy efficiency standards or establishes new ones for 15 types of appliances and equipment over the next several years.	1.8	0.50	11.45
Energy efficiency resource standard (EERS): Federal standards rise steadily to 20 percent for electricity and 10 percent for natural gas by 2030.	3.7	1.63	16.26
Building energy codes: New codes cut energy use in new residential and commercial buildings 15 percent annually until 2020, and 20 percent annually from 2021 to 2030.	1.2	2.12	14.19
Advanced buildings: An aggressive program ramps up and results in an additional 15 percent drop in energy use in new residential and commercial buildings by 2023 (beyond minimum building codes), with savings continuing at that level through 2030.	1.1	3.96	21.78
Research and development: Annual R&D investments reach $4.6 billion in 2030, and stimulate additional private-sector investments that reach $18.5 billion that year. These investments result in a 4.4 percent reduction in U.S. energy use by 2030.	1.8	4.65	18.50
Combined heat and power (CHP): A range of barrier-removing policies and annual investments in federal and state CHP programs lead to about 88,000 megawatts of new capacity by 2030—an average annual addition of 4,000 megawatts.	0.6	0.06	27.57
Industrial energy efficiency: Expanded federal programs, combined with local programs that support plant-level efforts, reduce industrial fuel use 10 percent (beyond that achieved by EERS and CHP) by 2030.	1.7	0.36	2.58
Rural energy efficiency: The federal government expands its farm bill Section 9006 technical assistance grants.	0.01	0.003	0.02
Petroleum feedstocks: Wider use of recycled feedstocks cuts industrial use of petroleum feedstocks 20 percent by 2030.	0.3	0.02	0.15
TOTAL	12.1	13.40	113.55

the economy.[39] The federal government abandoned many of those efforts in the early 1990s, when the price of electricity dropped and many states deregulated electricity markets. Only with the Farm Security and Rural Investment Act of 2002, known as the farm bill, did rural energy efficiency programs begin to reappear (Brown, Elliott, and Nadel 2005).

The Blueprint assumes that Section 9006 of the farm bill would continue. That section mandates annual grants of $35 million—including more than 40,000 individual grants—to provide technical assistance to farmers, to encourage them to rely on renewable energy and improve their energy efficiency. Under the Blueprint, such programs would enable farmers to cut their energy use 10–30 percent.

4.6.9. Use of Recycled Petroleum Feedstocks

The Blueprint builds on existing mandates for recycling plastics and other petrochemical products, and also assumes that research on using recycled materials in industrial processes would expand. The result is that the use of petroleum in industrial feedstocks drops a total of 12 percent by 2020, and 20 percent by 2030. These cuts are consistent with the impact of mandated plastic-recycling efforts in Germany (Elliott, Langer, and Nadel 2006).

4.7. The Bottom Line

Energy efficiency is the quickest, most cost-effective strategy for delivering significant and sustained cuts in carbon emissions. Innovative technologies and common-sense measures are available now, and can transform how our industries and buildings use energy over the next two decades (see Table 4.3). However, the nation needs to implement a suite of policies that builds on leading experiences at the state and federal level, to remove key barriers and stimulate investment. Once implemented, these policies can reduce total U.S. energy consumption 29 percent by 2030 while providing significant cost savings to consumers.

---

39 Because energy expenses account for up to 10 percent of a farm's budget, changes in energy costs can significantly affect the viability of operations in this low-profit-margin sector (Brown, Elliot, and Nadel 2005).

BOX 4.3.

The Many Faces of Energy Efficiency

Windows labeled "Low-E" keep buildings warmer in the winter and cooler in the summer. Energy Star labels help consumers identify the most energy-efficient products.

A blower-door test finds leaks that can be sealed, creating an airtight building with minimal heat and air-conditioning loss.

A properly sized HVAC system with centrally located ducts eliminates heat loss.

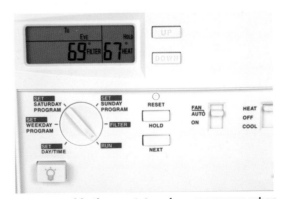

Programmable thermostats reduce energy use when residents are sleeping or not home.

Switchgrass field, Stephenville, TX

CHAPTER 5

Flipping the Switch to Cleaner Electricity

Electricity is an essential part of our daily lives and vital to our economy. It helps us light and cool our homes, refrigerate and cook our food, and wash and dry our clothes. Electricity also powers our offices, schools, hospitals, and factories. In fact, we have come to take its convenience for granted. We expect it to be there when we flip a switch—and at an affordable price.

Yet most people do not have a good understanding of where their electricity comes from, or of the impact our reliance on fossil fuels has on our climate, environment, public health, and public safety—and their significant hidden costs to our economy (see Figure 5.1).

The United States could greatly reduce its reliance on fossil fuels to generate electricity by moving to renewable resources such as wind, solar, geothermal, bioenergy, and hydropower. These homegrown energy sources are available in significant quantities across America, and we can deploy them quickly. They are also increasingly cost-effective in producing electricity, and they create jobs while reducing pollution.

As Chapter 4 noted, the nation has tremendous potential to reduce electricity use by improving the energy efficiency of our buildings and industries. However, expanding the use of renewable energy and other low-carbon technologies to generate electricity is also critical if we are to avoid the most dangerous effects of global warming.

The electricity sector was responsible for more than 40 percent of U.S. carbon dioxide emissions in 2007. Those emissions from power plants have grown by more than 33 percent since 1990—faster than heat-trapping emissions in any other sector of the economy, including transportation. And coal-burning power plants are the single largest source of carbon emissions, representing about one-third of the U.S. total—more than those from all our cars, SUVs, trucks, trains, and ships combined (EIA 2008d).

This chapter describes the current status and future prospects for using renewable energy and other low-

FIGURE 5.1. U.S. Electricity Generation by Source (2007)

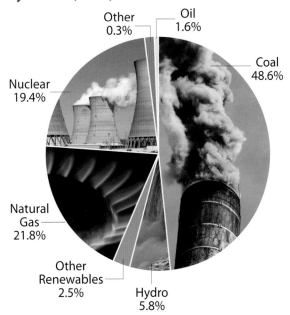

- Other 0.3%
- Oil 1.6%
- Coal 48.6%
- Nuclear 19.4%
- Natural Gas 21.8%
- Other Renewables 2.5%
- Hydro 5.8%

carbon technologies to provide a growing share of the nation's electricity needs. The chapter highlights key challenges to achieving widespread use of these technologies, and the public policies that can help us fulfill that goal.

5.1. Electricity from Renewable Energy Technologies

Diverse sources of renewable energy have the technical potential to provide all the electricity the nation needs many times over. Estimates of this potential consider the availability of strong winds, sunny skies, plant residues, heat from the earth, and fast-moving water throughout the United States, while accounting for some environmental and economic limits. However, such estimates do not consider conflicts over land use, the higher short-term costs of those resources, constraints on ramping up their use such as limits on transmission capacity, barriers to public acceptance, and

The United States currently generates nearly half of its electricity from coal, the most carbon-intensive energy source. Accelerating energy efficiency and the adoption of carbon-free renewable energy technologies such as wind and solar is needed to cut emissions and create savings from the electricity sector.

Renewable energy can create more jobs than fossil fuels because a larger share of renewable energy expenditures go to manufacturing, installation, and maintenance—all of which are typically more labor-intensive than mining and transporting fossil fuels. The U.S. wind industry created 35,000 jobs in 2008 alone.

other hurdles. Those factors will limit how quickly and to what extent the nation taps the full potential of renewable resources to produce electricity.

Several renewable energy technologies are available for widespread deployment today, or are projected to become commercially ready in the next two decades. In fact, in 2007 developers installed more than 8,600 megawatts of capacity for generating electricity from renewable sources (excluding conventional hydroelectric power)—topping new capacity from fossil fuels for the first time (EIA 2009a). And developers installed even more capacity to produce electricity from renewable sources in 2008. This section describes this recent progress as well as future prospects for the most promising renewable energy technologies.

5.1.1. Types of Renewable Technologies

5.1.1.1. Wind Power

Wind turbines convert the force of moving air into electricity. Like an airplane, the wind turns the blades using lift. Most modern wind turbines have three blades rotating around a horizontal axis. Smaller wind turbines used by homes, farms, and businesses range in size from a few hundred watts to 100 kilowatts or more. Larger wind turbines used for utility-scale generation range in size from about 500 kilowatts to more than three megawatts, have blades up to 52 meters long, and are mounted on towers up to 100 meters high.

Wind power is one of the most rapidly growing sources of electricity in the world—having increased by about 30 percent per year, on average, over the past decade (GWEC 2008). Developers installed more wind power over the past two years than in the previous 20. In 2008 the United States surpassed Germany to become the global leader in installed wind capacity, followed by Spain, India, and China. U.S. wind capacity grew by a record 5,250 megawatts in 2007, and 8,545 megawatts in 2008. This represented 42 percent of all new capacity for generating electricity in the country (AWEA 2009a).

As of March 2009, the United States had more than 28,000 megawatts of wind power capacity in 36 states (see Figure 5.2). Texas (7,900 megawatts) and Iowa (2,900 megawatts) have surpassed California (2,600 megawatts) to become the national leaders, followed by Minnesota, Washington, Colorado, Oregon, New York, and Kansas, which have more than 1,000 megawatts each (AWEA 2009b).

Wind power has been one of the bright spots in the struggling U.S. economy. According to the American

FIGURE 5.2. Installed Wind Power Capacity (2009)

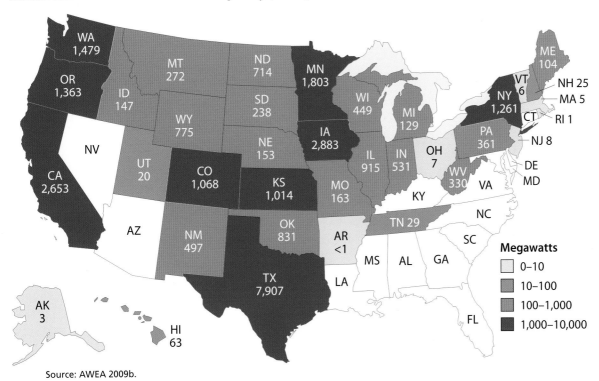

Source: AWEA 2009b.

Wind Energy Association, the industry now employs about 85,000 people, and added 35,000 new jobs last year alone. Developers invested some $27 billion in U.S. wind power over the past two years—much in agricultural and other rural areas. U.S. manufacturing of wind turbines and their components has also greatly expanded, with more than 70 new facilities opening, growing, or announced in 2007 and 2008. The industry estimates that these new facilities will create 13,000 high-paying jobs, and increase the share of domestically made components from about 30 percent in 2005 to 50 percent in 2008 (AWEA 2009b).

Other countries and several U.S. states are already relying on wind power to provide significant percentages of their electricity needs. In 2007, for example, wind power supplied more than 20 percent of electricity in Denmark, 12 percent in Spain, 9 percent in Portugal, 8 percent in Ireland, and 7 percent in Germany (Wiser and Bolinger 2008). Wind also provided an estimated 7.5 percent of electricity generated in-state in Minnesota and Iowa; 4–6 percent in Colorado, South Dakota, Oregon, and New Mexico; and 2–4 percent in 13 other states (Wiser and Bolinger 2008). Many of these states have committed to producing up to 25 percent of their electricity from wind and other renewable energy sources.

A comprehensive study by the U.S. Department of Energy (EERE 2008) found that wind power has the technical potential to provide more than 10 times today's U.S. electricity needs (see Table 5.1). That study also showed that expanding wind power from providing a little more than 1 percent of U.S. electricity in 2007 to 20 percent by 2030 is feasible, and would not affect the reliability of the nation's power supply. Achieving that target would require developing nearly 300,000 megawatts of new wind capacity, including 50,000 megawatts of offshore wind.

The DOE study found that, by 2030, that level of wind power would:

- Create more than 500,000 new U.S. jobs
- Displace 50 percent of the natural gas used to produce electricity, and reduce the use of coal by 18 percent, restraining rising fuel prices and stabilizing electricity rates
- Reduce global warming emissions from power plants by 825 million metric tons (20 percent)
- Reduce water use in the sector by 8 percent, saving 4 trillion gallons
- Cost 2 percent more than investing in new coal and natural gas plants—or 50 cents per month per household—including transmission costs but not federal incentives or any value for reducing carbon emissions

BOX 5.1.

SUCCESS STORY
The Little Country that Could

Children's tales don't often figure in grown-up discussions of energy policy, but Denmark's progress in tapping wind energy is reminiscent of *The Little Engine That Could.*

Denmark's story begins in 1973, the year OPEC (the Organization of Petroleum Exporting Countries) embargoed oil exports, creating debilitating shortages and skyrocketing prices. At that time Denmark relied on oil to produce 80 percent of its electricity. For the next few

> **Domestic investment in wind has made Denmark a global leader in turbine manufacturing. The industry accounts for roughly 20,000 jobs in Denmark—and 4 percent of its industrial production.**

years that country, much like the United States and other developed nations, invested in energy efficiency and alternative energy to prevent such a situation from occurring again.

When oil prices plummeted in the 1980s, however, the Danish and U.S. governments responded very differently. The United States stopped developing approaches to reducing its dependence on oil, but the Danish government continued to encourage the development of new energy sources and nascent technologies. Denmark reaps the benefits today as a net exporter of energy—a high percentage of which is carbon-free.

Denmark relied on a suite of policies to transform its economy into a much leaner, greener, and more secure one. Although it expanded development of conventional fuels off its coasts, Denmark focused principally on reducing demand for electricity and heat. The country stepped up its energy efficiency by insulating existing buildings, enacting stringent codes for new buildings and appliances, and relying on highly efficient combined-

Turbines located in the high-speed wind areas off the coast of Denmark generate large amounts of electricity.

heat-and-power plants to provide both electricity and heat. The primary power plant serving Copenhagen, for example, boasts an efficiency of more than 90 percent, compared with an average efficiency rate of 33 percent for a typical U.S. coal plant (Freese, Clemmer, and Nogee 2008).

Denmark fostered renewable energy as well, and today renewables supply 27 percent of the country's electricity—most of it from wind (Ministry of Climate and Energy 2008). With fewer than 70 wind turbines in 1980, the nation now has more than 5,000 providing 3,135 megawatts of capacity—enough to power more than 1.6 million typical American households (DWEA 2009).

Consistent, long-term policies encouraging the development of wind energy helped Denmark become a global leader. The government spurred investment in wind power by providing incentives that covered 30 percent of the costs of installing turbines until 1990. Denmark also required utilities to buy wind power at a fixed price until 1999. Although at that point the country required customers to pay any added costs of wind power, the government mandated that utilities provide 10-year fixed-rate contracts for wind developers, which helped them secure investment financing. Wind power also benefited from priority access to the electricity grid (GAO 2006).

This energy transformation helped Denmark expand its economy while reducing carbon emissions. Domestic investment in wind has made Denmark a global leader in turbine manufacturing. Vestas and Siemens Wind Power dominate global wind sales, and the industry accounts for roughly 20,000 jobs in Denmark—and 4 percent of its industrial production. While the economy has grown by roughly 75 percent in 25 years, energy consumption has remained stable, and the country has cut its carbon emissions in half since 1980 (Danish Energy Agency 2008; Ministry of Climate and Energy 2008).

Although Denmark is obviously much smaller than the United States, and its energy needs are much lower, the Danes have proved beyond a doubt that national foresightedness and perseverance—combined with smart policies and industrial innovation—can produce an extraordinary shift in a country's energy profile. The United States could learn much from the example of "the little country that could"—and did!

Growing interest in wind power is evident in the fact that at the end of 2007, developers of more than 225,000 megawatts of wind power capacity were seeking to connect with the transmission grid in 11 regions (Wiser and Bolinger 2008). This represents nine times the nation's installed wind capacity, roughly half of all generating capacity in transmission queues, and twice as much capacity as natural gas, the next-largest resource. Although many of these projects may not be built, many are in the planning phase.

While developers have so far sited all U.S. wind projects on land, they have shown considerable interest in developing offshore wind. At the end of 2007, seven U.S. states had seen active proposals for installing nearly 1,700 megawatts of offshore wind power (see Table 5.2). Developers are proposing to build most of these facilities off the Atlantic coast in the Northeast, close to population centers, where power is most needed. However, projects are also being considered off the Southeast and Texas coasts, and in the Great Lakes (Wiser and Bolinger 2008).

Wind power can provide an important economic boost to farmers. Large wind turbines typically use less than half an acre of land, including access roads, so farmers can continue to plant crops and graze livestock right up to the base of the turbines (as shown on this Trimont, MN, farm).

TABLE 5.1. Technical Potential for Producing U.S. Electricity from Renewable Sources

Renewable Resource	Electricity Generation Capacity Potential (gigawatts)	Electricity Generation (billion kilowatt-hours)	Renewable Electricity Generation as Percent of 2007 Electricity Use
Wind			
Land-Based	8,000	24,528	591%
Shallow Offshore	2,000	7,008	169%
Deep Offshore	3,000	11,826	285%
Subtotal	**13,000**	**43,362**	**1,044%**
Solar			
Distributed Photovoltaics	1,000	1,752	42%
Concentrating Solar Power	6,877	16,266	392%
Subtotal	**7,877**	**18,018**	**434%**
Bioenergy			
Energy Crops	83	584	14%
Agricultural Residues	114	801	19%
Forest Residues	33	231	6%
Urban Residues	15	104	3%
Landfill Gas	2.6	19	0.4%
Subtotal	**248**	**1,739**	**42%**
Geothermal			
Hydrothermal	33	260	6%
Enhanced Geothermal Systems	518	4,084	98%
Co-Produced with Oil and Gas	44	347	8%
Subtotal	**595**	**4,691**	**113%**
Hydropower			
Existing Conventional	77	259	6%
New Conventional	62	218	5%
Wave	90	260	6%
Hydrokinetic (tidal/in-stream)	53	140	3%
Subtotal	**283**	**888**	**21%**
Total	**22,000**	**68,659**	**1,653%**

Sources: See Appendix D online.

5.1.1.2. Solar Power

Our analysis included two main technologies for using solar power to supply electricity: photovoltaics (PV) and concentrating solar power (CSP). Both have been used to generate electricity for decades, though recent technological improvements and strong policy incentives have dramatically accelerated their growth. In 2007, global PV installations expanded by 62 percent from the previous year (Solarbuzz 2008). And after two decades of very little activity, the CSP market is also quickly gaining steam.

Photovoltaics, or solar cells, use semiconducting materials to convert direct sunlight to electricity. Most PV cells are made with silicon, the same material used to manufacture computer chips, although manufacturers are using new materials to make some PV cells. PV cells are often used in rooftop solar energy systems, and to power remote, off-grid applications. However, power producers have also recently shown interest in developing multi-megawatt PV projects that would connect to the transmission grid.

CSP typically works by concentrating direct sunlight on a fluid-filled receiver. This heated fluid then drives a turbine to produce electricity. CSP is most often used in large, utility-scale plants that are far from urban areas yet connected to the transmission grid. Most existing CSP plants rely on curved (parabolic) mirrors to focus solar radiation. However, a number of companies are developing large CSP plants that use "power towers" to collect solar energy from ground-mounted heliostats—or slightly curved mirrors—and concentrate solar radiation on distributed receivers.

The technical potential of U.S. solar power is huge. PV panels installed on less than 1 percent of the U.S. land

TABLE 5.2. Proposed U.S. Offshore Wind Projects (2007)

State	Proposed Capacity (megawatts)
Massachusetts	783
New Jersey	350
Delaware	200
New York	160
Texas	150
Ohio	20
Rhode Island	20
Georgia	10
Total	**1,693 MW**

Note: The 450 megawatt project in Delaware was reduced to 200 megawatts and a 20 megawatt project in Rhode Island was added.

Source: Wiser and Bolinger 2008.

area could generate the equivalent of the country's entire annual electricity needs, as could CSP plants covering a 100-square-mile area.

The southwestern United States—with its arid deserts and minimal cloud cover—is home to some of the world's best solar resources. The National Renewable Energy Laboratory (NREL) estimates that CSP has the potential to generate 7,000 gigawatts of electricity in the Southwest—after screening out urban centers, national parks, other protected areas, and lands with slopes greater than 1 percent (SETP 2007). This potential is roughly 10 times the nation's entire current capacity to generate electricity. NREL also identified optimal locations for 200 gigawatts of CSP, taking into account proximity to existing transmission lines, and estimated that the nation could build as much as 80 gigawatts of CSP capacity by 2030 (see Figure 5.3).

Although the United States lags behind other countries in tapping CSP, the industry is poised for significant growth because of new state and federal policies. In the Economic Stimulus Package of October 2008 Congress extended the 30 percent investment tax credit for solar energy projects for eight years. Several states have also adopted renewable electricity standards and financial incentives to expand the share of solar in their electricity mix. And several utilities have signed contracts to develop

...omes that use solar panels and energy efficiency are not only sustainable but also affordable. The Make It Right project (www.makeitrightnola.org), which is helping to rebuild New Orleans homes destroyed by Hurricane Katrina, showcases designs that put low electricity bills within everyone's reach.

BOX 5.2.

SUCCESS STORY
Surprises in the Desert

Deserts have long been imagined as hot and desolate landscapes—but their reputations have been burnished recently. Deserts are now more likely to be appreciated as unique and often surprisingly diverse environments. Approximately 40 miles southeast of Las Vegas, the desert does indeed hold a most surprising find: a power plant generating electricity from the sun.

When most people think of solar energy, images of photovoltaic panels on rooftops come to mind. But there is another kind: concentrated solar power (CSP), which uses mirrors to collect and transform the heat of the sun into steam, which spins a generator. CSP's relatively simple approach enables it to produce renewable electricity on a scale comparable to conventional coal and natural gas plants.

The third largest solar power plant in the world—and the largest CSP plant in the United States—was built outside Boulder City, NV, in June 2007. The Nevada Solar One plant uses 760 long, tubular mirrors (or parabolic troughs) to concentrate the sun's energy on solar receivers.[40] The receivers heat a mineral oil fluid to 734°F, which turns water into steam that powers a turbine to generate electricity. The solar receivers track the sun's movement, allowing the facility to produce electricity during all of the hours in which the sun is brightest.

The solar fields themselves occupy an area roughly the size of 200 football fields. The plant's maximum capacity is 75 megawatts, and it generates about 134 million kilowatt-hours of electricity each year—enough to power the lights, appliances, and electronics in 14,000 average U.S. homes. This near-zero-carbon electricity reduces global warming emissions by an amount equivalent to taking 20,000 cars off the road each year.

CSP is now sparking a lot of attention. Interest is especially high in the desert Southwest, which contains large open spaces and some of the world's best solar resources. This area is also close to some of the country's largest and fastest-growing population centers. As of July 2008, the federal Bureau of Land Management

The world's third largest solar power plant sits in the Nevada desert, generating clean electricity.

had received 125 applications to develop large-scale solar facilities on public lands (EIA 2008). In California alone, developers have proposed more than 3,500 megawatts of CSP projects, which are now under regulatory review (CEC 2008a).

Another piece of good news is that the construction of CSP plants creates good jobs. Estimates suggest that every 100 megawatts of installed CSP capacity creates 455 temporary construction jobs (Stoddard, Abiecunas, and O'Connell 2006). The Nevada One facility, for example, provided over 800 construction jobs for about 17 months, and now permanently employs approximately 30 people (ACCIONA 2009).

As with any renewable energy technology, CSP must be built in an environmentally responsible manner. Because many CSP projects are sited in desert areas, developers must avoid disrupting the natural habitats of unique desert plants and animals, and minimize the water used for cooling. But if careful policies guide environmentally responsible CSP development, our deserts may continue to be surprising places—where catclaw acacia and solar power plants alike delight the occasional visitor.

---

40   The Nevada plant is owned by ACCIONA Solar Power, a subsidiary of ACCIONA Energy. Headquartered in Madrid, Spain, this energy company develops and manages renewable energy plants and infrastructure projects throughout the world.

FIGURE 5.3. The Potential of Concentrating Solar Power

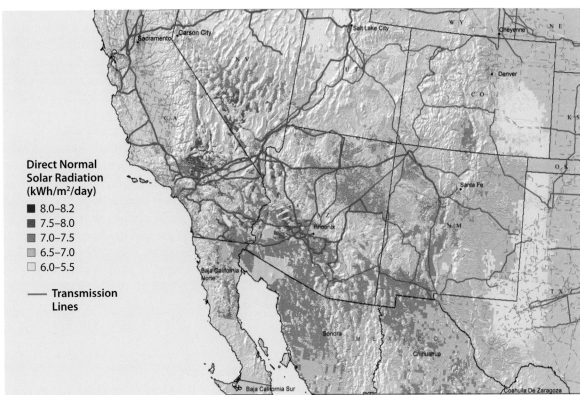

Note: Potentially sensitive environmental lands, major urban areas, water features, areas with slope > 3%, and remaining areas less than 1 sq. km were excluded to identify those areas with the greatest potential for development.

Source: NREL, July 2007.

both distributed and large-scale solar projects. For example, utilities in California and Arizona have contracted for enough new CSP projects to more than triple existing global capacity.

The U.S. solar energy industry employs more than 80,000 people and created more than 15,000 jobs in the last two years. One recent study estimates that the industry will create 440,000 permanent jobs and spur $325 billion in private investment by 2016, given the federal investment tax credit (Navigant 2008).

5.1.1.3. Geothermal Energy

Geothermal energy—heat from the earth—can be used to heat and cool buildings directly, or to produce electricity in power plants. Almost all existing geothermal power plants use hot water and steam from hydrothermal reservoirs in the earth's crust to drive electric generators. These plants rely on holes drilled into the rock to more effectively capture the hot water and steam. Much like power plants that run on coal and natural gas, geothermal plants can supply electricity around the clock.

More than 8,900 megawatts of geothermal capacity in 24 countries now produce enough electricity to meet the annual needs of nearly 12 million typical U.S. households (GEA 2008a). Geothermal plants produce 25 percent or more of the electricity produced in the Philippines, Iceland, and El Salvador. The United States has more geothermal capacity than any other country, with nearly 3,000 megawatts in seven western states. About two-thirds of this capacity is in California, where 43 geothermal plants provide nearly 5 percent of the state's electricity (CEC 2008).

While geothermal now provides only 0.4 percent of U.S. electricity, it has the potential to play a much larger role—thereby reducing carbon emissions and moving the nation toward a cleaner, more sustainable energy system. In its first comprehensive assessment in more than 30 years, the U.S. Geological Survey (USGS) estimated that conventional hydrothermal sources on private and accessible public lands across 13 western states have the potential capacity to produce 8,000–73,000 megawatts, with a mean estimate of 33,000 megawatts (Williams et al. 2008). State and federal policies are likely to spur developers to tap some of this potential in the next few years. The Geothermal Energy Association estimates that 103 projects now under development

FIGURE 5.4. How Advanced Geothermal Systems Work

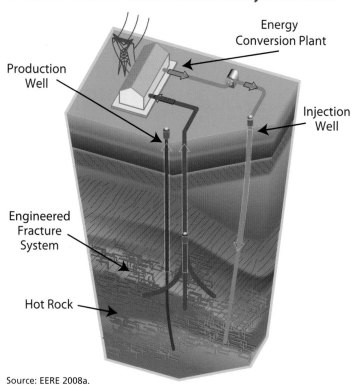

Source: EERE 2008a.

Enhanced geothermal systems tap into hot rock at greater depths than conventional geothermal systems—approaching the depths of oil and gas wells—to expand the economically recoverable amount of heat and power stored under Earth's surface. The Department of Energy, several universities, the geothermal industry, and venture capital firms are collaborating on research and demonstration projects to harness the potential of this technology.

in the West could provide up to 3,960 megawatts of new capacity (GEA 2008b).

While most near-term capacity will likely come from hydrothermal sources, the USGS study also found that enhanced geothermal systems (EGS) could provide another 345,100–727,900 megawatts of capacity, with a mean estimate of 517,800 megwatts (see Table 5.1). That means this resource could supply nearly all of today's U.S. electricity needs (Williams et al. 2008).

EGS entails engineering hydrothermal reservoirs in hot rocks that are typically at greater depths below the earth's surface than conventional sources. Developers do this by drilling production wells and pumping high-pressure water through the rocks to break them up. The plants then pump more water through the broken hot rocks, where it heats up, returns to the surface as steam, and powers turbines to generate electricity (see Figure 5.4). Finally, the water is returned to the reservoir through injection wells to complete the circulation loop. Plants that use a closed-loop binary cycle release

no fluids or heat-trapping emissions other than water vapor, which may be used for cooling (EERE 2008a).

The DOE, several universities, the geothermal industry, and venture capital firms are collaborating on research and demonstration projects to harness the potential of EGS. Google.org is playing an especially active role in promoting the technology (Google 2008). Australia, France, Germany, and Japan also have R&D programs to make EGS commercially viable.

One of the goals of these efforts is to expand the economically recoverable resource to depths approaching those used in oil and gas drilling. Depths of six kilometers (19,685 feet) have enough heat to make geothermal energy viable in many more areas (see Figure 5.5). The oil and gas industry has already successfully drilled to such depths. Shell Oil holds the record, having drilled to a depth of more than 10 kilometers (33,200 feet) in the Gulf of Mexico in January 2004 (GEA 2008c).

The Blueprint analysis includes both hydrothermal and EGS technologies.

5.1.1.4. Biopower

Biomass is the oldest source of renewable energy, coming into use when our ancestors learned the secret of fire. Humans have been burning biomass to make heat, steam, and electricity ever since.

The Blueprint analysis considers a wide variety of bioenergy resources. These include lower-cost biomass residues from forests, crops, urban areas, the forest products industry, and landfill gas, which is mostly

In the near term, the most cost-effective biopower options for reducing the global warming emissions associated with electricity generation are 1) co-firing biomass with coal, 2) biomass combined heat and power, and 3) landfill gas. The Grayling power station in Michigan uses waste wood from local sawmills and the forest products industry.

FIGURE 5.5. Geothermal Potential

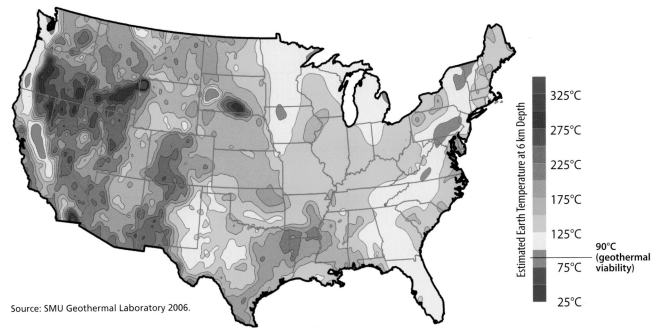

Source: SMU Geothermal Laboratory 2006.

methane from decomposing organic matter. The analysis also includes crops grown primarily for use in producing energy, such as fast-growing poplars and switchgrass (a native prairie grass). The availability and quantity of these resources varies from region to region based on many factors, including climate, soils, geography, and population. (For more information, see Appendix G online.)

The Blueprint includes three main approaches to large-scale production of electricity from biomass: dedicated biomass power plants, which run solely on biomass; coal plants that burn biomass along with coal; and the use of biomass to produce both electricity and steam—also known as combined heat and power, or CHP—in the forest products and biofuels industries. Our analysis also includes electricity production from landfill gas, which is a fairly limited resource in the United States. According to the U.S. Environmental Protection Agency (EPA), 469 landfill gas-to-electricity projects with 1,440 megawatts of capacity are now operating, while another 520 landfills with 1,200 megawatts of potential capacity could be developed (EPA 2008d).

In the short term, co-firing of biomass with coal, CHP, and landfill gas are likely to be the most cost-effective uses of biomass to generate power. However, dedicated biomass gasification plants—a technology that is similar to advanced coal gasification plants (see below)—could make a contribution in the next two decades.

Biomass supplied more than 50 percent of U.S. electricity generated from renewable sources other than hydro in 2007. More than 10,000 megawatts of biomass capacity produced about 1.3 percent of the nation's electricity that year. Biomass also provides 20 percent of total CHP capacity in the industrial sector—nearly all in the forest products industry (EIA 2008e).

The growth of biopower will depend on the availability of resources, land-use and harvesting practices, and the amount of biomass used to make fuel for transportation and other uses. Analysts have produced widely varying estimates of the potential for electricity from biomass.

For example, a 2005 DOE study found that the nation has the technical potential to produce more than a billion tons of biomass for energy use (Perlack et al. 2005). If all of that was used to produce electricity, it could have met more than 40 percent of our electricity needs in 2007 (see Table 5.1).

In a study of the implementation of a 25 percent renewable electricity standard by 2025, the Energy Information Administration (EIA) assumed that 598 million tons of biomass would be available, and that it could meet 12 percent of the nation's electricity needs by 2025 (EIA 2007). In another study, NREL estimated that more than 423 million metric tons of biomass would be available each year (see Figure 5.6) (ASES 2007).

In our analysis, we assumed that only 367 million tons of biomass would be available to produce both

FIGURE 5.6. Bioenergy Potential

The National Renewable Energy Laboratory estimates that the United States has the potential to produce 423 million metric tons of biomass each year from agricultural, forest, and urban waste. This map shows where this resource is distributed throughout the country—the Midwest, Southeast, and West Coast have the greatest bioenergy potential. Energy crops such as switchgrass, which are not represented on the map, have the potential to supply up to 240 million metric tons of additional biomass per year.

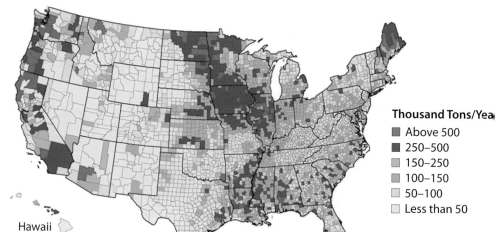

Thousand Tons/Year

■ Above 500
■ 250–500
▨ 150–250
▨ 100–150
□ 50–100
□ Less than 50

Source: NREL 2005; Milbrandt 2005.

electricity and biofuels. That conservative estimate accounts for potential land-use conflicts, and tries to ensure the sustainable production and use of the biomass.

To minimize the impact of growing energy crops on land now used to grow food crops, we excluded 50 percent of the switchgrass supply assumed by the EIA. That allows for most switchgrass to grow on pasture and marginal agricultural lands—and also provides much greater cuts in carbon emissions (for more details, see Appendix G online). The potential contribution of biomass to electricity production in our analysis is therefore just one-third of that identified in the DOE study, and 60 percent of that in the EIA study.

5.1.1.5. Hydropower

Harnessing the kinetic energy in moving water is one of the oldest ways to generate electricity. The most common approach is to dam free-flowing rivers and then use gravity to force the water through turbines to produce electricity.

The United States produced about 6 percent of its electricity supply from conventional hydropower sources in 2007. While environmental concerns limit the potential for new projects, the nation can expand its conventional hydropower by adding and upgrading turbines at existing facilities, and by adding turbines to dams that do not now generate power, with minimal environmental impact.

The Blueprint case estimates that such incremental hydro projects have the potential to produce about 5

percent of today's U.S. electricity needs. Our analysis does not include new technologies that can harness the kinetic energy from currents in undammed rivers, tides, oceans, and constructed waterways, because the NEMS model does not represent those resources. Those technologies have the potential to supply more than 140 gigawatts of new capacity, and thus could provide 9 percent of the nation's current electricity use (Dixon and Bedard 2007).

5.1.2. The Vast Potential of Electricity from Renewable Sources

Potential The major renewable energy technologies (wind, solar, geothermal, bioenergy, and hydropower) together have the technical potential to generate more than 16 times the amount of electricity the nation now needs (see Table 5.1). In fact, wind, solar, and geothermal each have the potential to meet today's electricity needs. Of course, economic, physical, and other limitations mean that the nation will not tap all this potential.

Still, several recent studies have shown that renewable energy can provide a significant share of future electricity needs, even after accounting for many of these factors. For example, the American Solar Energy Society (ASES)—working with experts at NREL—projected that the United States could obtain virtually all the cuts in carbon emissions it needs by 2030 by aggressively pursuing both energy efficiency and electricity from renewable energy (ASES 2007). After accounting for efficiency improvements, the study found that a diverse mix of renewable energy technologies could provide about 50 percent of the remaining U.S. electricity needs by 2030.

A follow-up analysis found that the savings on energy bills from energy efficiency would more than offset the estimated $30 billion that renewable energy would cost under this scenario. The result would be net savings of more than $80 billion per year (Kutscher 2008). That study might well have underestimated the resulting cuts in heat-trapping emissions, because it did not consider all the options for producing electricity from renewable sources, or technologies for storing electricity other than solar thermal.

More than 20 comprehensive analyses over the past decade have found that using renewable sources to provide up to 25 percent of U.S. electricity needs is both achievable and affordable (Nogee, Deyette, and Clemmer 2007). For example, a 2009 Union of Concerned Scientists study—using the same modified version of the EIA's NEMS model that we used for the Blueprint—found that a national renewable electricity standard of 25 percent by 2025 would lower electricity and natural gas bills in all 50 states, by reducing demand for fossil fuels and increasing competition among power producers (UCS 2009). Cumulative national savings to consumers and businesses would total $95 billion by 2030.

A 2009 EIA study arrived at similar conclusions, despite using more pessimistic assumptions about the viability of renewable energy technologies. That study projected that a renewable electricity standard of 25 percent by 2025 would lower consumer natural gas bills slightly—offsetting slightly higher electricity bills (EIA 2009b). By 2030, the impact on consumers' cumulative electricity and natural gas bills under two different scenarios would range from a small cost of $8.4 billion (0.2 percent) to a slight savings of $2.5 bil-

lion (0.1 percent). Similarly, a 2007 EIA study of a 25 percent by 2025 renewable electricity standard found $2 billion in cumulative savings on combined electricity and natural gas bills through 2030 (EIA 2007).

These studies have also shown that renewable energy can make a significant contribution to U.S. electricity needs while maintaining the reliability of the nation's electricity supply. The EIA and UCS analyses project that renewable technologies that operate around the clock—such as biomass, geothermal, landfill gas, and incremental hydroelectric plants—would generate 33–66 percent of the nation's electricity under a national renewable electricity standard.

Regional systems for transmitting electricity could easily integrate the remaining power produced from wind and solar at a very modest cost, and without storing the power. Studies by U.S. and European utilities have found that reliance on wind energy for as much as 25 percent of electricity needs would add no more than five dollars per megawatt-hour—or less than 10 percent—in grid integration costs to the wholesale cost of wind (Holttinen et al. 2007).

5.1.3. Costs of Producing Electricity from Renewable Sources

Costs

An analysis by NREL shows that the costs of wind, solar, and geothermal technologies fell by 50–90 percent between 1980 and 2005 (see Figure 5.7). The main drivers of these drops were advances in technology, and growing volumes and economies of scale in manufacturing, building, and operating these plants—spurred by government policies and funding for R&D.

Despite these important gains, the costs of most renewable and conventional energy technologies rose over the past few years. Figure 5.7 does not reflect these increases, which are primarily due to the escalating costs

The United States has more geothermal capacity than any other country, with nearly 3,000 megawatts in seven western states. Projects like the Geysers in California are harnessing only a small fraction of a much larger U.S. potential.

FIGURE 5.7. Declining Cost of Renewable Electricity

(levelized cost of electricity, in cents per kilowatt-hour)

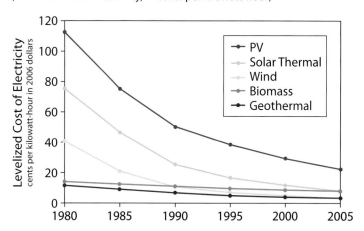

Source: NREL Energy Analysis Office. See *www.nrel.gov/analysis/docs/cost_curves_2005.ppt.*

Note: The levelized cost of electricity includes the annualized costs of capital and operation and maintenance. This graph reflects historical trends, not precise data on annual costs.

come competitive with other options for producing electricity.

The costs of emerging renewable technologies, such as solar PV, concentrating solar thermal, and offshore wind, are projected to decline significantly over time because wind and solar are modular and can be mass-produced to drive down costs. Advanced fossil fuel and nuclear plants are large-scale, and thus likely to see more modest cost reductions through more standardized designs and engineering.

5.1.4. Key Challenges for Producing Electricity from Renewable Sources

5.1.4.1. Siting

Renewable energy technologies allow the nation to avoid or greatly reduce many of the environmental and public health effects from mining and transporting fuels and producing electricity from fossil fuels and nuclear power. However, despite these important benefits, care must be taken in siting renewable energy projects to minimize potential environmental impacts.

For example, while studies show that wind power usually results in far fewer bird deaths than other causes, a few wind projects have seen significant numbers of birds and bats colliding with the turbines (Erickson et al. 2001). Siting geothermal, large-scale solar, and offshore wind, wave, and tidal projects can also be challenging because many of the best sites are on federally controlled lands and seas, and often require both federal and state approval. Obtaining the required approvals and leases can often take several years, which can deter investors.

Efforts are under way to minimize these impacts as the industry expands, through careful planning, site selection, research, and monitoring. Efforts are also under way to streamline the approval process and improve cooperation between local, state, and federal agencies while ensuring responsible development.

5.1.4.2. Ensuring the Sustainability of Bioenergy and Wise Land Use

When grown and used sustainably, biomass produces almost no net carbon emissions. If biopower used some form of carbon capture and storage (CCS), the technology could actually lower the concentration of carbon in the atmosphere. However, unsustainable biomass harvesting practices can alter the amount of carbon stored and released by soils and trees, and the production of biomass can sometimes require the use of fossil fuels. The overall impact on global warming emissions of generating electricity from biomass

of materials, labor, and fuel; the weak dollar; and bottlenecks in the supply chain.

The recent economic downturn and corresponding declines in the price of fuel and materials—combined with a significant increase in U.S. manufacturing of renewable energy technologies (primarily wind and solar)—is already reversing these trends. NREL and many other experts project that the costs of renewable energy will follow the historic trend because of continued growth in the industry and advances in the technology. Stable, long-term national policies that help eliminate market barriers and encourage the growth of renewable energy will likely accelerate these declining costs.

Under these conditions—along with a national policy that puts a price on carbon emissions—renewable energy technologies will become increasingly cost-effective compared with new coal, natural gas, and nuclear power plants. In fact, some renewable technologies, such as wind and geothermal at sites with high-quality resources, are competitive with new coal and natural gas plants without incentives or a price on carbon emissions (see Figure 5.8).

Advanced coal and natural gas plants with carbon capture and storage, and advanced nuclear plants, in contrast (see below), are more expensive than conventional coal and natural gas plants and many renewable energy technologies, even when a cost of $40 per ton of carbon emissions is included. These technologies will need to drop significantly in cost to be-

FIGURE 5.8. Cost of Electricity from Various Sources (2015)

(levelized cost of electricity, in 2006 dollars per megawatt-hour)

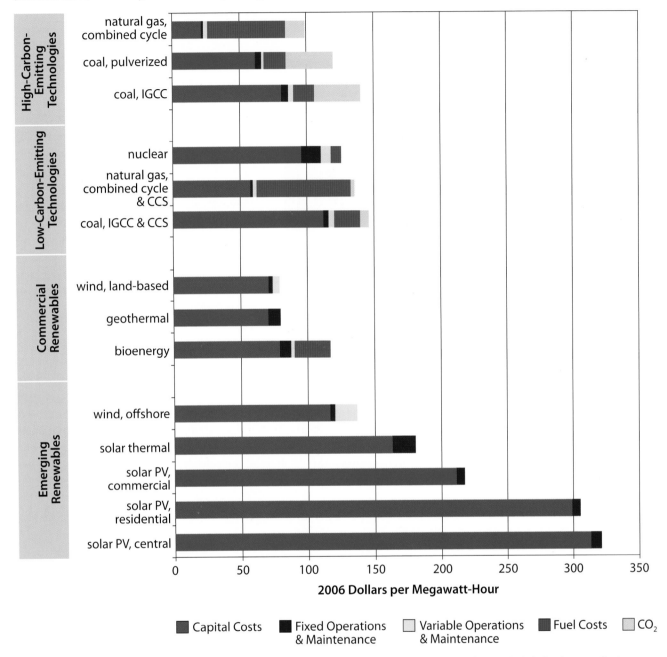

Note: IGCC = integrated gasification combined cycle; CCS = carbon capture and storage. The levelized cost of electricity includes the annualized cost of capital, operation and maintenance, and fuel from the Reference case, as well as a CO_2 price of $40/ton for illustrative purposes (where applicable). It does not include the cost of transmitting power or integrating facilities into the grid, or cost reductions from tax credits and other incentives for renewable and conventional technologies reflected in the model. See Appendix D for more details on technology cost assumptions and Appendix A for more details on fuel prices (both available online).

When grown and used sustainably, biomass produces almost no net carbon emissions. If biopower used some form of carbon capture and storage (CCS), the technology could actually lower the concentration of carbon in the atmosphere.

BOX 5.3.

Technologies on the Horizon: Renewable Energy

A researcher testing nanotechnology for solar panels.

Biopower. Biomass gasification with carbon capture and storage (CCS) is a promising technology that could reduce the net amount of carbon in the atmosphere. If grown and used sustainably, biomass absorbs CO_2 from the atmosphere, which could then be captured during the gasification process and sequestered in geologic formations.

Several companies are also working on using algae to produce energy, and to store—or sequester—carbon. One company has completed a demonstration project using algae to sequester flue gases from a coal power plant, and is considering recycling the biomass into the host facility for use as a fuel.

Geothermal. An MIT study estimated that the United States has the potential to develop 44,000 megawatts of geothermal capacity by 2050 by co-producing electricity, oil, and natural gas at oil and gas fields—primarily in the Southeast and Southern Plains (Tester et al. 2006). The study projects that such advanced geothermal systems could supply 10 percent of U.S. baseload electricity by that year, given R&D and deployment over the next 10 years.

Our analysis did not include several renewable energy technologies that are at an early stage of development, but that offer promise over the long-term (after 2030). Our analysis also did not include some technologies that could make a contribution over the next two decades, but that our model was unable to adequately represent. These technologies include:

Solar. Thin-film PV cells offer promising new applications for solar energy, such as in roof tiles and building facades. While such cells are less costly to produce than semiconductor-grade crystalline-silicon wafers, they typically have much lower efficiencies. Still, venture capitalists had invested more than $600 million in thin-film PV by 2008, and the technology is projected to account for 25 percent of the PV market and $26 billion in sales by 2013 (Miller 2008).

Researchers and several companies are also exploring the use of solar nanotechnology: thin films of microscopic particles and tiny semiconducting crystals that release conducting electrons after absorbing light. Nanotechnology could revolutionize the solar industry by making solar cells cheaper, more efficient, lighter, and easier to install.

Hydrokinetic. New technologies that harness the hydrokinetic energy in currents in undammed rivers, tides, oceans, and constructed waterways could provide more than 140 gigawatts of new electrical capacity—enough to power more than 67 million U.S. homes (Dixon and Bedard 2007).

Renewable energy technologies for heating and cooling. These technologies are commercially available today but supply only 2–3 percent of worldwide demand. Mature technologies include solar, biomass, and geothermal heating and cooling systems. Use of these technologies is growing rapidly in the European Union, where strong policies promoting renewal energy are helping to offset higher up-front costs (IEA 2007).

Advanced storage. These technologies would allow renewable but variable energy sources—such as wind, solar, and hydrokinetic energy—to meet electricity needs around the clock. The most promising storage options now seeing targeted R&D include compressed air storage, reversible-flow batteries, thermal storage, and pumped hydro. These technologies could bring many benefits to operators of electricity grids, including greater stability of power, better management of peak demand and transmission capacity, and higher-quality power (Peters and O'Malley 2008).

A tidal turbine in New York City's East River.

depends on the type of biomass, the method of producing and delivering it, the energy source being displaced, and alternative uses for the resource.

It is also important to consider potential carbon emissions created by changes in land use. Some forms of biomass—such as native perennials grown on land that would not be used for food, and biomass from waste products such as agricultural residues—do not change the way we use our land, and can therefore significantly reduce global warming emissions. However, changing the way we use land to produce biomass for energy may indirectly affect land use in other countries. For example, turning forested land that is high in stored carbon into cropland to compensate for shrinking cropland in the United States may mean that biomass creates more carbon emissions than it prevents.

Fragmented jurisdiction over the existing transmission system allows any single state to effectively veto the construction of new multistate transmission lines by refusing to grant the needed permits.

5.1.4.3. Expanding the Transmission Grid

A lack of capacity for transmitting renewable electricity from remote areas to urban areas is another key challenge. While most renewable energy technologies can be deployed quickly, obtaining approvals to site new transmission lines and actually building them typically takes several years. While new transmission lines are often controversial, the public is beginning to show a greater willingness to accept them if they are carrying power from clean renewable sources instead of high-carbon fossil fuels and nuclear power.

Fragmented jurisdiction over the existing transmission system allows any single state to effectively veto the construction of new multistate transmission lines by refusing to grant the needed permits. Federal land-use agencies also lack a consistent policy for siting transmission lines. To address those challenges, the nation needs a new federal siting authority to integrate state and regional processes for approving new transmission lines, and to help plan for and integrate new renewable resources and distributed power plants into the grid,

New investments in transmission capacity will be needed to move electricity from areas rich in renewable resources to areas where the electricity is actually used. To ensure the most efficient transmission possible, these investments should include improvements to the transmission grid, changes in the process for building new lines, and innovative methods for financing new lines.

while taking into account options for managing demand. Such an authority should also allocate costs fairly among all users of the transmission system, and ensure the protection of sensitive environmental and cultural resources.

Several renewable energy technologies could share transmission lines. In fact, combining bioenergy, geothermal, landfill gas, and hydro projects—which provide baseload power—with wind and solar projects, which provide varying amounts of power, can allow more cost-effective use of new transmission lines and upgrades. State, regional, and national agencies are now considering how to increase the capacity of the grid to transmit power from "renewable energy zones" to areas of high demand, to capture some of these benefits. In

the future, technologies for storing electricity, creating a smart grid, and forecasting wind resources will further improve the use of transmission lines and help integrate wind and solar projects into the grid.

Policies 5.1.5. Key Policies for Increasing Electricity from Renewable Sources

We examined a package of market-oriented policies needed to overcome the market barriers that now limit growth of renewable energy, to spur investment by consumers and the power producers. This package included both standards and incentives, as no single policy can address the range of market barriers faced by renewable energy technologies that are at different stages of development.

5.1.5.1. Renewable Electricity Standard

The renewable electricity standard (RES)—also known as a renewable portfolio standard—has emerged as a popular and effective tool for reducing market barriers and stimulating new markets for renewable energy (UCS 2007). The RES is a flexible, market-based policy that requires electricity providers to gradually increase the amount of renewable energy in the power they supply. By using a system of tradable credits for compliance, the RES encourages competition among all renewable energy sources, rewarding the lowest-cost technologies and creating an incentive to drive down costs.

As of January 2009, 28 states and the District of Columbia have adopted an RES.[41] Our Reference case includes the renewable energy that has resulted from these policies.

The Blueprint includes a national RES that begins at 4 percent of projected electricity sales in 2010, and ramps up gradually to 40 percent in 2030—after accounting for the cuts in demand for electricity resulting from improvements in energy efficiency. This represents about 25 percent of electricity sales in the Reference case in 2030, not including energy efficiency. The ramp-up rate of 1–1.5 percent of electricity sales annually in the Reference case (without efficiency) is consistent with standards in leading states such as Illinois, Minnesota, New Jersey, and Oregon, as well as the stronger national RES proposals.[42]

---

41 For detailed information on state renewable electricity standards, see *http://www.ucsusa.org/res*.

42 Reps. Markey (D-MA) and Platts (R-PA) have introduced a national RES of 25 percent by 2025 in the House, while Sens. Udall (D-CO), Udall (D-NM), and Klobuchar (D-MN) have introduced similar proposals in the Senate. President Obama also supported a 25 percent RES during his campaign.

The Blueprint also assumed that:
- All U.S. electricity providers must meet the targets
- Eligible technologies include biomass, geothermal, incremental or new capacity at existing hydroelectric facilities, landfill gas, solar, and wind
- Providers can use existing renewable energy sources, except existing hydro, to meet the targets

> **Experts agree that deploying enough renewable energy resources to achieve strong targets for cutting carbon emissions will be impossible unless the nation dramatically modernizes and expands the grid for transmitting electricity.**

5.1.5.2. Tax Credits

Production and investment tax credits help defray the typically higher up-front costs of renewable energy technologies. Such credits also help level the playing field with fossil and nuclear technologies, which have historically received much higher tax subsidies (Goldberg 2000; Sissine 1994).

Both the Reference case and the Blueprint case include the extension and expansion of tax credits for renewable energy technologies that were part of the 2008 Economic Stimulus Package. That legislation includes a one-year extension (through 2009) of the production tax credit for wind; a two-year extension (through 2010) of the production tax credit for geothermal, solar, biomass, landfill gas, and certain hydro facilities; and an eight-year extension (through 2016) of the 30 percent investment tax credit for solar and small wind systems. Our analysis did not include the tax credits and incentives from the American Recovery and Reinvestment Act of 2009, because it was enacted after we had completed our modeling.

5.1.5.3. Other Renewable Energy Policies

We also recommend several other policies to help commercialize a broad range of renewable energy technologies. While our analysis did not explicitly model those

policies, we assumed that they would help facilitate the development of the technologies that the analysis did include, as well as help providers meet the national renewable electricity targets. These policies include:

Greening our transmission system. Experts agree that deploying enough renewable energy resources to achieve strong targets for cutting carbon emissions will be impossible unless the nation dramatically modernizes and expands the grid for transmitting electricity. Addressing this problem quickly will require reforming the management and operation of the grid, creating new mechanisms for financing and recovering the costs of an expanded grid, and creating processes for siting new transmission lines. These measures will help producers of electricity generated from carbon-free renewable resources connect to the grid. Coupled with these efforts must be initiatives that encourage energy efficiency, demand-side management, and smart grid improvements, while discouraging access to new lines from high-carbon emitters.

Our analysis assumed that new national policies will facilitate new transmission lines and upgrades of existing lines to enable power producers to meet national renewable electricity targets. While we did not explicitly model these policies, we did include the costs of building new transmission lines for new renewable, fossil-fueled, and nuclear power plants, and we allocated those costs to all electricity users based on EIA assumptions. The Blueprint analysis also included the costs of siting and connecting wind projects, and

New Jersey policies promoting clean energy helped finance the nation's largest single-roof solar project (at the Atlantic City Convention Center). Unveiled in March 2009, the project meets 26 percent of the building's electrical needs and avoids the release of more than 2,300 tons of CO_2 annually.

transmitting the power they produce, as the use of wind grows, based on an analysis by NREL for the EIA.

More funding for R&D. More funding for research and development is essential for commercializing electricity based on renewable energy, as well as other low-carbon technologies. R&D drives innovation and performance gains while helping to lower the cost of emerging technologies. Our analysis assumed that federal R&D funding for renewable energy would double over a five-year period.

Net metering offers consumers who generate their own electricity (via a rooftop solar panel, small wind turbine, or other eligible technologies) a credit on their electricity bills for excess power they feed into the electrical grid.

Net metering. Net metering allows consumers who generate their own electricity from renewable technologies—such as a rooftop solar panel or a small wind turbine—to feed excess power back into the electricity system and spin their meter backward. Forty-one states and the District of Columbia now have net metering requirements. Adopting this policy at the national level would encourage the development of small wind, solar, biomass, and geothermal systems for producing electricity.

Feed-in (or fixed-price) tariffs. Feed-in tariffs provide a specific, guaranteed price for electricity from renewable energy sources—typically over a 10–20-year period. European countries such as Germany have long had such tariffs, and they are gaining momentum among the states, primarily to promote small-scale and community-owned power projects. State feed-in tariffs targeted at smaller, higher-cost emerging technologies and locally owned projects

would complement renewable electricity standards, as those tend to benefit larger, lower-cost projects and technologies that are closer to commercialization.

Financial incentives. Financial incentives such as rebates, grants, and loans can stimulate investment and help bring renewable energy technologies to market. Funding for such programs can come from various state sources, such as renewable energy funds, and federal sources such as clean renewable energy bonds (CREBS), which Congress recently extended through 2009 in the Economic Stimulus Package.

5.2. Electricity from Fossil Fuels with Carbon Capture and Storage

While renewable energy technologies have the technical potential to produce all the nation's electricity and eliminate carbon emissions from that sector, the country must address many challenges to realize that potential. Given the uncertainties in our ability to surmount those market barriers, and to guarantee advances in renewable technologies and reductions in their cost, the nation may need other low-carbon approaches to avoid the most dangerous effects of global warming.

Carbon capture and storage (CCS) is an emerging technology that could allow electricity producers to capture carbon dioxide from power plants and pump it into underground formations, where it would ideally remain safely stored over the very long term (see Figure 5.9). This approach is being investigated today primarily to reduce carbon emissions from coal-fired power plants. However, it could also be used to prevent emissions from natural-gas-fired power plants or other industrial facilities that release a significant stream of carbon dioxide. And facilities that burn or gasify biomass could actually provide carbon-negative power—that is, they could store carbon dioxide recently removed from the atmosphere through the photosynthesis of the plants they use as fuel—if they relied on CCS.

5.2.1. Types of CCS Technologies

One CCS technology is pre-combustion capture, which can be used with integrated gasification combined-cycle (IGCC) coal plants. IGCC plants heat the coal to create a synthetic gas, or syngas. The syngas fuels a combustion turbine used to generate electricity, and the waste heat from that process creates additional power via a steam turbine. Converting the coal into a gas allows operators to remove CO_2 before combustion, when it is in a more concentrated and pressurized form.

IGCC is a relatively new technology: only four plants now operate worldwide, although developers

A relatively new technology called integrated gasification combined cycle (IGCC) has the potential to capture carbon emissions more easily than traditional coal plants. Only two IGCC coal plants are operating in the United States today, including this one near Tampa, FL; neither is currently capturing carbon.

have announced several others. Interest in IGCC is strong because it is seen as more amenable to carbon capture than traditional coal plants, though none of the IGCC coal plants now operating employ CCS.

Another capture technology under development is post-combustion capture, which would be used with traditional coal plants. Collecting CO_2 after combustion is more challenging because the gas is more diluted, requiring greater energy to collect and compress it. One way to collect the CO_2 is with amine scrubbers, now used to capture CO_2 in much smaller industrial applications. Another approach, called oxy-fueling, would fuel a coal plant with oxygen rather than background air, yielding a purer stream of CO_2 after combustion. Oxy-fueling is in an earlier stage of investigation than the other capture methods.

Our analysis included only pre-combustion carbon capture in new coal IGCC and natural gas combined-cycle plants, because NEMS currently does not have the capacity to model post-combustion capture technologies.

Both pre- and post-combustion technologies are expected to capture 85–95 percent of a coal plant's carbon emissions. When factoring in the fuel used to power the CO_2 capture process, though, the actual rate of carbon emissions avoided per unit of electricity is expected to fall to 80–90 percent (IPCC 2005).

Researchers are investigating underground storage of CO_2—often called sequestration—in several projects around the world. Options for storing the CO_2 include pumping it into depleted oil or gas fields, coal seams that cannot be mined, and deep saline aquifers. Detailed analyses of CCS have concluded that long-term geologic storage of CO_2 is technically feasible, though careful site selection is critical (MIT 2007; IPCC 2005).

While many components of CCS are in use in other, usually smaller, applications and pilot projects, there have not yet been any commercial-scale, fully integrated projects demonstrating CCS at coal-fired power plants. Developers have announced several such projects, including in the United States, though most are seeking more government funding before moving forward.

5.2.2. Potential of Carbon Capture and Storage

Potential

Some 500 coal plants provided half the nation's electricity in 2007—and produced about one-third of all U.S. carbon emissions. A typical new coal plant averages about 600 megawatts in size. The DOE estimates that geologic formations in North America have the capacity to store hundreds of years' worth of U.S. carbon emissions, based on today's rate. However, some areas are far from suitable storage formations (NETL 2006).

Computer models cited by the Intergovernmental Panel on Climate Change indicate that CCS could eventually contribute 15–54 percent of the cuts in carbon emissions needed by 2100. Recent government studies of proposed U.S. climate legislation also show large-scale development of advanced coal plants with CCS before 2030 (EIA 2008; EPA 2008a). Studies further show that CCS deployment could significantly lower the cost of stabilizing concentrations of heat-trapping gases in the atmosphere (Creyts et al. 2007; EPRI 2007; MIT 2007; IPCC 2005).

However, all these studies use optimistic assumptions about capital costs, ramp-up rates, and the ability to scale up the enormous infrastructure needed to transport, store, and monitor the emissions. Government studies also include generous incentives for CCS in proposed federal legislation, which tip the balance toward CCS versus other technologies. Studies that do not include these incentives, and that use more reasonable assumptions about capital costs and ramp-up rates, show advanced coal with CCS making a much smaller contribution by 2030 (e.g., EPRI 2008).

5.2.3. Costs of Carbon Capture and Storage

Costs

The DOE estimates that adding post-combustion capture (using amine scrubbing) to a traditional coal plant would increase the cost of electricity 81–85 percent. Adding pre-combustion capture to an IGCC plant would raise the cost of electricity 32–40 percent, but the underlying IGCC plant costs more than a traditional coal plant. These estimates suggest that IGCC plants with pre-combustion CCS would cost somewhat less than traditional plants with post-combustion CCS. However, MIT analysts contend that it is too soon to know which technology would cost less (MIT 2007; NETL 2007).

The higher cost of energy in these approaches reflects both the higher capital costs of adding CCS and the resulting losses in the plant's output. Post-combustion capture is particularly energy intensive: amine scrubbing is expected to reduce a plant's power output by a quarter or more, even if engineers integrate CCS into the plant's original design. If CCS is added as a retrofit, the energy penalty and higher cost of energy would be much greater.

Because no one has yet built a coal-fired power plant with CCS, estimates of the technology's performance and cost are more uncertain than those of other approaches to cutting global warming emissions.

5.2.4. Key Challenges for Carbon Capture and Storage

Challenges

CCS faces many challenges. For the technology to play a major role in reducing heat-trapping emissions, the nation would need an enormous new infrastructure to capture, process, transport (usually by pipeline), and store large quantities of CO_2. For

FIGURE 5.9. How Carbon Capture and Storage Works

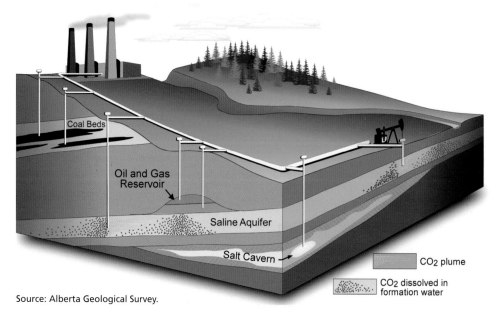

Coal Beds

Oil and Gas Reservoir

Saline Aquifer

Salt Cavern

CO_2 plume

CO_2 dissolved in formation water

Source: Alberta Geological Survey.

Carbon capture and storage (CCS) technology would allow the CO_2 from coal-fired power plants to be captured and injected into geologic formations such as depleted oil and gas reservoirs, unmineable coal seams, or saline aquifers. No coal-fired power plants currently employ this technology, but several commercial-scale demonstration projects have been announced around the world.

To make responsible energy choices we must consider the full costs of how we generate electricity. For example, conventional coal technology presents threats beyond the power plant—mountaintop removal mining has irreversibly damaged Appalachian mountains and buried more than 700 miles of biologically diverse streams.

example, if 60 percent of the CO_2 now released by U.S. coal plants were captured and stored, the volume would equal that of all U.S. oil consumption (MIT 2007).

Environmental concerns linked to CCS include the risk that CO_2 will leak back into the atmosphere. Slow leaks would contribute to global warming, while fast leaks could pose a local danger, as high concentrations of CO_2 are fatal. Another concern is that CO_2 could migrate in unexpected ways, picking up toxic components underground and contaminating freshwater aquifers. The risk of leakage and migration rises in the presence of abandoned oil and gas wells, which can provide conduits for the CO_2.

Reducing these risks will require careful site selection and long-term monitoring, which in turn will require the development and enforcement of rigorous regulations. Long-term liability questions must also be answered.

CCS added to coal plants will also do nothing to reduce the serious environmental and social costs of mining and transporting coal. Indeed, coal plants with CCS will require more coal per megawatt-hour of electricity they produce than plants without it, given that the capture process consumes energy. And while some of the other air pollutants from today's coal plants would likely decline if they were redesigned to employ

CCS, other environmental effects such as water use could increase or stay the same.

One unique environmental benefit of CCS is its potential to be paired with biomass to produce electricity that actually reduces atmospheric concentrations—not just emissions—of carbon. As plants grow, they absorb CO_2 from the atmosphere. The CCS process—used at a facility that gasifies or burns biomass—would then turn the atmospheric carbon captured by the plants into geologic carbon. Such carbon-negative energy facilities could play an important role in fighting global warming in the decades ahead.

5.2.5. Key Policies for Carbon Capture and Storage

Policies

In *Coal Power in a Warming World: A Sensible Transition to Cleaner Energy Options,* UCS analysts conclude that CCS has enough potential to play a significant role in reducing carbon emissions to warrant further investigation and investment, despite its many challenges (Freese, Clemmer, and Nogee 2008). The nation needs to reduce the one-third of U.S. carbon emissions that come from coal-based electricity, and to stop building new coal plants without CCS technology. UCS therefore supports federal funding for 5 to 10 demonstration projects of various

According to the Climate 2030 Blueprint, advanced nuclear power plants will not become cost-competitive with other low-carbon energy sources before 2030. Such plants could play a role in reducing carbon emissions after 2030, or sooner if their costs are reduced more quickly than expected, but nuclear power still carries substantial safety and security risks—with long-term consequences—that must be addressed before we build a new generation of plants.

types, to help determine the technology's true costs and effectiveness.

The Blueprint reflects this financial support by assuming that the nation would build eight new IGCC plants with CCS, funded by a small portion of the revenues from auctioning carbon allowances under a cap-and-trade program. The analysis assumes that all the CCS projects would be new IGCC plants because NEMS does not have the ability to model other types of CCS projects.

Both the Reference and Blueprint cases also include the 30 percent investment tax credit for advanced coal and CCS projects, up to a maximum of $2.55 billion, in the October 2008 Emergency Economic Stabilization Act. That legislation also provides an incentive of $10–20 per ton of CO_2 for the use of CCS in enhanced oil recovery and in other geologic formations.

Because it includes an economywide cap-and-trade program that puts a price on carbon emissions, the Blueprint provides an incentive to reduce emissions from existing coal plants and develop new plants with CCS. While not explicitly modeled in our analysis, a

CO_2 performance standard would prevent the construction of new coal plants unless and until they can employ CCS in their original design. As *Coal Power in a Warming World* also notes, the nation needs new statutes and stronger regulations to reduce the environmental and social costs of coal use—from mining through waste disposal—that will accompany any funding or other policy support for CCS.

5.3. Electricity from Advanced Nuclear Plants

A nuclear power plant generates electricity by splitting uranium atoms in a controlled fission process. The fission reaction creates heat, which is used to make steam, which turns a turbine (as in most other electricity plants). Two types of reactors—boiling water reactors (BWRs) and pressurized water reactors (PWRs)—are in use in the United States today (UCS 2003).

Nuclear power plants could play a role in reducing global warming emissions, because they emit almost no carbon when they operate. Other parts of the nuclear fuel cycle emit carbon dioxide, especially today's uranium enrichment processes, which rely on coal-fired power plants and inefficient technology. However, some studies have found those emissions to be roughly comparable to those from manufacturing and installing wind power and hydropower facilities (UCS 2003).

The United States now obtains about 20 percent of its electricity from 104 nuclear power plants (EIA 2008). Thanks to better operating performance, the "capacity factor" of U.S. nuclear reactors rose from 56 percent in 1980 to 91.5 percent in 2007 (EIA 2008). However, U.S. utilities ordered no new nuclear plants after 1978, and canceled all plants ordered after 1973. Other countries have continued to build nuclear plants, although at a much slower rate than during the peak years of the 1970s and 1980s.

The Nuclear Regulatory Commission (NRC) is in the process of extending the licenses for most, if not all, U.S. plants now operating—from an original 40-year period to 60 years. Almost all these plants would have to be retired and decommissioned between 2030 and 2050, unless the NRC extends their licenses again. However, the economic and technical feasibility of doing so has not been established.

5.3.1. Types of Advanced Nuclear Technologies

Fourteen companies have submitted applications to the NRC to build and operate 26 plants at 17 sites, although no utility has actually ordered a new plant yet.[43] These applications reference five plant designs—of which the NRC has certified only two. And one of those, the AP1000, has undergone significant design changes since it was certified

The five designs offer evolutionary improvements on existing plants: they are somewhat simpler, relying more on "passive" safety systems and less on pumps and valves. The industry and the NRC had hoped that these upgrades—along with a streamlined licensing process and greater standardization—would improve the safety of nuclear power plants and reduce their costs. However, the goal of standardization has so far proved elusive and the licensing process has not yet been fully tested.

Of all the new reactor designs under serious consideration for use in the United States, only one—the Evolutionary Power Reactor (EPR)—appears to have the potential to be significantly safer and more secure than existing reactors, provided that it is built to the stricter safety standards required by France and Ger-

many. However, because the EPR design does not feature the same safety shortcuts as the passive designs, including the AP1000, Standard & Poor's rated it as the most risky with regard to capital costs.

Several companies are also working on much smaller plants in the 10–150-megawatt range, compared with 1,000–1,600 megawatts for traditional designs. By making modular units and siting them underground, these companies hope to rely on mass production to achieve economies of scale and improve safety and security. However, no power companies have submitted such designs to the NRC for licensing, so we cannot yet evaluate the companies' claims.

Other new designs in research and development—known as Generation IV designs—aim to achieve major leaps in safety and cost. However, a significant number of engineering problems remain to be solved, so we cannot yet evaluate the claims for Generation IV plants either. In fact, they are not expected to be ready for deployment before 2030. Because the Blueprint analysis examined costs and benefits through 2030 only, we did not include these advanced designs.

5.3.2. Potential of Advanced Nuclear Power

Potential

According to the International Atomic Energy Agency, the world has enough uranium supplies to fuel the existing 400 nuclear plants for more than 100 years, and to expand that fleet by 38–80 percent by 2030 (IAEA 2008). Some proponents argue that the reprocessing of used nuclear fuel to extract plutonium could create a virtually unlimited supply of fuel for use in "fast breeder" reactors. However, reprocessing is many times more expensive than the traditional "once-through" fuel cycle. Reprocessing also greatly increases the risk that weapons-usable nuclear materials will be diverted—as well as the volume of radioactive wastes requiring disposal (UCS 2007a). While uneconomical today, some scientists believe that seawater could eventually supply virtually unlimited quantities of uranium at lower cost than fuel made from reprocessing (Garwin 2001).

Nuclear power could therefore, in theory, contribute to a climate-friendly future. A recent EIA analysis of the impact of climate legislation projected as much as 268 gigawatts of new nuclear capacity by 2030—supplying 58 percent of total U.S. demand, and a significant share of the needed cuts in carbon emissions

---

43 AmerenUE recently announced that it has canceled plans to build the proposed Callaway 2 reactor in Missouri.

FIGURE 5.10. Nuclear Power Plant Construction Costs Rising Faster than Other Technologies

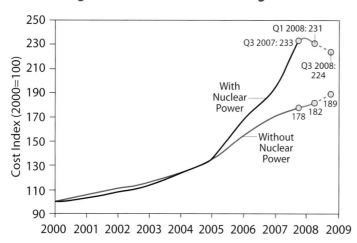

Source: CERA 2008.

Over the past nine years, construction costs have risen for all electricity-generating facilities. But over the past four years, the costs for nuclear power plants have risen much more dramatically than other technologies. Between 2000 and the third quarter of 2008, construction costs have risen 89 percent for all non-nuclear technologies, compared with 124 percent when nuclear power is included.

(EIA 2008). However, the EIA assumed very outdated and low "overnight" construction costs of $2,475 per kilowatt (in 2006 dollars)—well below current industry estimates. (Overnight costs do not include financing or escalating costs during construction.) The industry also faces significant constraints to deploying new nuclear plants that rapidly, and to that extent (see below).

5.3.3. Costs of Advanced Nuclear Power

Costs

The cost of electricity from nuclear power plants is largely driven by the cost of constructing them. The fuel and operating costs of existing nuclear plants are generally lower than those of other conventional technologies for producing electricity. However, very high construction costs—stemming from long construction periods and associated financing costs—have been the economic Achilles heel of the nuclear industry.

During the 1970s and 1980s, with cost overruns averaging more than 200 percent, utilities abandoned more than half of the planned nuclear fleet during construction. And the plants they did complete usually led to significant increases in electricity rates. The total losses to ratepayers, taxpayers, and shareholders stemming from cost overruns, canceled plants, and stranded costs well exceed $300 billion in today's dollars (Schlissel, Mullet, and Alvarez 2009).

Reliably projecting construction costs for new U.S. nuclear plants is impossible, because the nation has no recent experience to draw upon. Recent experience with reactors under construction in Europe, however—along with recent trends in the overall cost of commodities and construction—show the same vulnerability to cost escalation that plagued the last generation of nuclear plants. Only three years after its 2005 groundbreaking, for example, the Olkiluoto plant in Finland was reportedly three years behind schedule, with cost overruns topping 50 percent. The project has encountered numerous quality problems, and the principals are in arbitration over responsibility for the overruns (*The Guardian* 2009).

Construction costs have risen over the past five years for all technologies used to produce electricity—but most dramatically for nuclear plants—as shown in Figure 5.10 (CERA 2008). For example, in November 2008, Duke Energy revised its estimate of overnight construction costs for two nuclear units proposed for Cherokee County, SC, to $5,000 per kilowatt. Several other analysts and developers of nuclear plants have estimated a range of $3,800–$5,500 per kilowatt. Utilities applying for loan guarantees in November 2008 estimated that the costs of their proposed 21 plants—including cost escalation and the cost of financing—would total $188 billion, an average of $9 billion per plant, or more than $6,700 per kilowatt.

Our analysis assumed that overnight capital costs for new nuclear plants would initially average $4,400 per kilowatt for those with a 2016 in-service date, not including financing costs. The NEMS model calculates the cost with financing to be $6,900 per kilowatt, which is close to the average estimates available when we finalized assumptions for our model. Our figure is lower because we assumed that industry learning would reduce costs by nearly 7 percent by 2030—or half the rate projected by the EIA based on international experience.

France and South Korea have achieved higher learning rates largely because of standardization: one company builds one plant design over and over. In the fractured U.S. industry, with 17 companies proposing to build 26 units based on five different designs (with more on the horizon), high learning rates are optimistic. Indeed, the U.S. nuclear industry saw construction costs rise steadily through almost the entire last generation of plants (EIA 1986), making any future cost reductions through learning very uncertain. Continued cost escalation would be more consistent with the U.S. experience.

5.3.4. Key Challenges for Advanced Nuclear Power

Challenges

Nuclear technologies pose a number of unique and complex challenges. An expansion of nuclear power would increase the risks to human safety and security (UCS 2007a). These include a release of radiation because of a reactor meltdown or terrorist attack. If proposals for reprocessing nuclear waste move forward, the detonation of a nuclear weapon made with materials from a civilian nuclear power system could produce massive civilian deaths. Such an incident would obviously also threaten the viability of nuclear power.

After 50 years of nuclear power, a mix of technical and political challenges has meant that no country has yet licensed a long-term nuclear waste repository. The proposed Yucca Mountain site in the United States has been plagued with technical, managerial, and political problems (GAO 2006), and the Obama administration announced in early 2009 that it would no longer pursue it as a permanent repository. While nuclear waste can be stored safely in hardened concrete casks on-site or in a central repository in the short run, successfully licensing long-term storage is a critical challenge for the industry to see substantial growth.

Nuclear plants also require enormous volumes of water for cooling. In both Europe and the United States, nuclear plants have had to reduce power output or shut down during some drought periods (GAO 2006). Water requirements—especially as global warming leads to more drought conditions in some regions—could limit the expansion of nuclear power.

Nuclear power is sometimes touted as a "domestic" energy resource, although the United States imports about 80 percent of its nuclear fuel. These imports come primarily from stable and friendly countries: Canada, Australia, and South Africa. However, nuclear power will displace little if any imported oil from less stable and potentially less friendly regions, because the United States produces very little electricity from oil today. Furthermore, overseas corporations such as AREVA, a French-based company, and Mitsubishi Heavy Industries, which is based in Japan, will make most major components for nuclear plants.

Siting and permitting nuclear facilities present other significant challenges. While many surveys have shown growing public acceptance of nuclear power during the last few years, people still generally rank it lower than all other sources of electricity except perhaps coal. The NRC has significantly streamlined its process for licensing nuclear power plants to limit opportunities for interest groups to challenge them, but this process has yet to be tested.

While nuclear plants may make a significant long-term contribution to reducing U.S. carbon emissions, they are unlikely to do so before at least 2030. Beyond the challenges just noted, the nation would have to rebuild its civilian nuclear infrastructure, which has been in decline for two to three decades.

For example, nuclear engineering programs in the United States have declined by half since the mid-1970s, and only 80 companies are qualified to produce nuclear-grade materials, down from 400 two decades ago. Most important, only two manufacturing facilities in the world are capable of making heavy components for nuclear plants, such as reactor pressure vessels—although AREVA announced its intention to build a vessel in Virginia with Northrop Grumman Corp.

As a result, the Organization for Economic Cooperation and Development has estimated that the industry can produce an average of only 12 plants per year worldwide until about 2030, rising to 54 plants per year from 2030 to 2050 (OECD 2008). Although the United States represents about one-quarter of global energy use and carbon emissions, it is unlikely that developers would install more than three or four U.S. plants per year before 2030.[44]

Scaling up the nuclear industry to make a long-term contribution along the lines suggested by MIT analysts—1,000 to 1,500 new 1,000-megawatt plants worldwide, with 300 in the United States—would require the construction of 11 to 22 new enrichment facilities, as well as a new Yucca Mountain-sized waste repository somewhere in the world every four years (MIT 2003, The Keystone Center 2007). These facilities would pose great challenges for preventing proliferation of radioactive materials that could be used for weapons—as well as for siting those facilities. However, given the pressing need for cuts in carbon emissions of 80 percent or more by mid-century, the nuclear power option should not be off the table. Instead, it should receive R&D funding aimed at resolving these critical challenges.

---

44 Installing more than that amount in the United States could actually worsen global warming, by diverting reactors from countries such as China that use more coal and have higher rates of carbon emissions.

The industry hopes to make a number of advanced reactor designs—referred to as Generation IV—available sometime after 2025 to 2030. These designs aim to achieve much higher safety levels and lower costs. The industry faces numerous challenges in meeting those goals, however, and we cannot meaningfully evaluate the prospects that it will do so at this time (UCS 2007a). In any case, such reactors are not expected to be commercially available until after the time period we analyzed.

BOX 5.4.

Key Assumptions for Technologies Used to Produce Electricity

- **Escalation of construction costs.** We included recent increases in construction and commodity costs for all technologies, based on data from actual projects, input from experts, and power plant cost indices. We assumed that the costs of all technologies continue to rise 2.5 percent per year (after accounting for inflation) until 2015.

- **Wind.** We included land-based, offshore, and small wind technologies. We based our capital costs on a large sample of actual projects from a database at Lawrence Berkeley National Laboratory (LBNL). We used an analysis from the National Renewable Energy Laboratory (NREL), conducted for the EIA, to develop regional wind supply curves that include added costs for siting, transmitting, and integrating wind power as its use grows. We also assumed increases in wind capacity factors (a measure of power production) and a 10 percent reduction in capital costs by 2030 from technological learning, based on assumptions from a report from the DOE on producing 20 percent of U.S. electricity from wind power by 2030 (EERE 2008).

- **Solar.** We assumed expanded use of concentrating solar power (CSP) and distributed (small-scale) and utility-scale photovoltaics through 2020, based on actual proposals. We also assumed faster learning for solar photovoltaics, to match the EIA's assumptions for other emerging technologies. We assumed that the amount of heat that CSP can store to produce electricity during periods of high demand rises over time.

- **Bioenergy.** Key technologies included burning biomass along with coal in existing coal plants, dedicated biomass gasification plants, the use of biomass to produce combined heat and power in the industrial sector, and the use of methane gas from landfills.

- **Geothermal.** We included a supply curve for hydrothermal and enhanced geothermal systems in the West, developed by NREL and other experts. This supply curve incorporates recent increases in the costs of exploring potential sites, drilling, and building geothermal power plants.

- **Hydropower.** We assumed incremental amounts of hydropower from upgrades and new capacity at existing dams, and counted both new sources of power as contributing to a national standard for renewable electricity.

- **Carbon capture and storage.** We included this as an option for advanced coal gasification and natural gas combined-cycle plants, with costs and performance based on recent studies and proposed projects.

- **Nuclear.** We assumed that existing plants are relicensed and continue to operate through their 20-year license extension, and that they are then retired, as the EIA also assumes. We based assumptions on the costs and performance of new advanced plants primarily on recent project proposals and studies.

- **Transmission.** We included the costs of new capacity for transmitting electricity for all renewable, fossil, and nuclear technologies. We also added costs for the growing amounts of wind power, based on the NREL analysis conducted for the EIA.

(See Appendix D online for more details.)

5.3.5. Key Policies for Advanced Nuclear Power

Policies

Both the Reference and Blueprint cases include existing incentives and policy support for the next generation of nuclear power plants. For example, both cases include the existing production tax credit of 1.8 cents per kilowatt-hour (adjusted annually for inflation) for new nuclear plants that begin operation by 2020. The credit is available for the first eight years of operation, and is limited to $125 million per gigawatt of capacity annually, up to 6 gigawatts of total new capacity. However, if more than 6 gigawatts are under construction by January 1, 2014, those plants can share in the credits.

Both the Reference and Blueprint cases also include up to $18.5 billion in incentives available through the DOE's current loan guarantee program. In October 2008, the DOE received applications from 17 companies to build 21 new reactors at 14 nuclear plants. Those projects—which would provide a total of 28,800 megawatts of capacity—would qualify for $122 billion in loan guarantees. Because not enough funding is available for all the projects, and because the details of each one are unavailable, we simply assumed that the loan guarantees will spur the development of 4,400 megawatts of new nuclear capacity by 2030 ($18.5 billion divided by $122 billion times 28,800 megawatts).

The Blueprint's economywide cap-and-trade policy would provide an additional incentive to build new nuclear plants rather than coal and natural gas plants, because owners of the latter would have to buy allowances to emit carbon. The Blueprint case does not assume any additional policy support for advanced nuclear plants.

BOX 5.5.

Key Assumptions for Electricity Policies

POLICIES IN THE REFERENCE CASE

- **State renewable electricity standards.** These specify the amount of electricity that power suppliers must obtain from renewable energy sources. We replaced the EIA's estimate with our own projections for state standards through 2030. We applied those projections to the 28 states—plus Washington, DC—with such standards as of November 2008.

- **Tax credits.** We included the tax credit extensions for renewable energy and advanced fossil fuel technologies that were part of the Economic Stimulus Package (H.R. 6049) passed by Congress in October 2008.

- **Nuclear loan guarantees.** We assumed that the $18.5 billion in loan guarantees spur the construction of four new nuclear plants with 4,400 megawatts of capacity by 2020, based on applications received by the U.S. Department of Energy in October 2008.

ADDITIONAL POLICIES IN THE BLUEPRINT

- **Efficiency.** Policies to increase energy efficiency in buildings and industry (see Chapter 4) reduce electricity demand 35 percent by 2030 compared with the Reference case.

- **Combined heat and power (CHP).** Policies and incentives to increase the use of natural gas combined-heat-and-power systems in industry and commercial buildings (see Chapter 4) enable this technology to provide 16 percent of U.S. electricity generation by 2030.

- **National renewable electricity standard.** This standard requires retail electricity providers to obtain 40 percent of remaining electricity demand (after reductions for efficiency improvements and CHP) from renewable energy (wind, solar, geothermal, bioenergy, and incremental hydropower) by 2030.

- **Coal with carbon capture and storage (CCS) demonstration program.** This new federal program provides $9 billion to cover the incremental costs of adding CCS at eight new, full-scale advanced coal plants—known as integrated gasification combined-cycle plants, which turn coal into gas—from 2013 to 2016 in several regions.

Downtown Arlington, VA

CHAPTER 6

You Can Get There from Here: Transportation

America's transportation system is intricately woven into our daily lives. The most obvious examples are the light-duty vehicles (cars, SUVs, pickups, and minivans) we use to get to work, do errands, or visit family and friends. We rely on trucks, trains, and ships to move goods as well, and on garbage trucks to haul our waste. We also spend a lot of time on airplanes, while some people use public transit, walk, or bike.

All this travel and shipping add up when it comes to global warming. The production and use of fuel for transportation in the United States is directly responsible for more than 2 billion metric tons of carbon dioxide and other heat-trapping emissions annually.[45] That puts the U.S. transportation system second to power plants as the biggest contributors to the nation's global warming emissions—producing about 30 percent of the total, and more than one-third (36 percent) of all carbon dioxide emissions.

The biggest transportation sources of heat-trapping emissions are light-duty vehicles, which account for more than 60 percent of transportation's total, and about one-fifth of the nation's total (see Figure 6.1). Next in line are medium- and heavy-duty vehicles at 18 percent, followed by air at 10 pcercent, and then shipping, rail, military, and other uses.

The impact of these vehicles also adds up when it comes to America's addiction to oil. Transportation depends on petroleum for 98 percent of its fuel, and accounts for more than two-thirds of all petroleum products used in the United States (CTA 2008). In 2007, with average gasoline prices at more than $2.75 per gallon, Americans spent more than $575 billion on transportation fuels such as gasoline, diesel, and jet fuel (EIA 2008a).

FIGURE 6.1. The Sources of Transportation Heat-Trapping Emissions (2005)

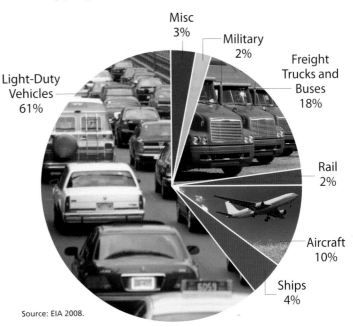

Misc 3%
Military 2%
Freight Trucks and Buses 18%
Light-Duty Vehicles 61%
Rail 2%
Aircraft 10%
Ships 4%

Source: EIA 2008.

Much of the oil used to make those fuels is imported, so transportation demand shipped nearly $200 billion out of our economy and into the hands of oil-exporting nations (EIA 2008a).

The impact of transportation's near-exclusive use of petroleum products goes beyond carbon emissions and the cost of fuel. The past 40 years have brought five significant spikes in oil prices—every one soon followed by an economic recession (CTA 2008). Although analysts have tied the most recent recession to problems with the housing and credit markets, spiking oil and gasoline prices likely had a significant impact, given these historical trends.

---

45   That is more than the amount produced from burning fossil fuels in all sectors in every nation except China and the United States (Marland, Boden, and Andres 2008).

A surprisingly small percentage of the fuel in your car's gas tank is actually used to move you down the road—most of the energy contained in fuel is wasted on inefficiencies in the engine and transmission, or lost while idling in stop-and-go traffic. Increasing fuel economy will help reduce these inefficiencies and decrease heat-trapping carbon emissions (which currently average about 1.25 pounds per mile driven, based on an average fuel economy of 20 miles per gallon for new cars and light trucks).

Three key ingredients for a more stable transportation system. The transportation sector is so large that no single solution will end our oil addiction and cut global warming emissions as much as we need. But the lack of a silver bullet does not mean there are no solutions. It simply means we need to take advantage of a variety of options to address these challenges.

The easiest way to think of this opportunity is to break it into three parts:

- Tapping technology to improve the efficiency of vehicles and their air conditioning systems.
- Shifting away from oil toward cleaner alternatives.
- Providing smarter transportation options to cut down on the number of miles we spend stuck in traffic in our cars.

As with a table or a stool, strengthening these three legs can provide both a stable climate and a secure energy future.

6.1. Driving Change: Technologies to Improve Fuel Efficiency and Air Conditioning

Only 15–20 percent of the energy in our fuel tanks actually goes to propelling today's cars and light trucks down the road. Most is effectively thrown away because of inefficiencies in the engine and transmission systems, or is wasted when we are stuck at a traffic light with the engine running or in stop-and-go traffic. Some of this energy also powers air conditioning, fans,

lights, and the growing use of onboard electronics. Of the fuel that is not simply wasted, most is needed to push air out of the way, keep tires rolling, and speed up the 1.5 to 3 tons of metal, plastic, and glass in our cars and trucks.

That, in a nutshell, is why new cars and light trucks sold in 2005 averaged only about 25 miles per gallon on government tests and about 20 mpg on the road—about the same as they did two decades ago (OTAQ 2008). As a result, the average new vehicle is responsible for nearly one and a quarter pounds of carbon dioxide and other heat-trapping emissions for every mile it is driven (one pound from the tailpipe, and another quarter-pound from making and delivering the fuel) (ANL 2008). The average auto—driven about 11,500 miles annually—is responsible for about 6.5 metric tons of global warming emissions every year.

The bigger trucks used to ship goods and haul garbage waste less fuel because they tend to have more efficient engines and transmissions. However, these vehicles carry a lot more weight, and their boxy shapes mean they must push much more air out of the way. Bigger trucks also use heavy-duty tires, which also waste more energy when they roll.

As a result, medium-duty trucks, such as those used to deliver packages, average 8–8.5 mpg of gasoline equivalent, while the heaviest trucks, such as 18-wheelers, average only about 6 mpg of gasoline equivalent (EIA 2008a). This low fuel economy means that medium-duty trucks are responsible for about three pounds of global warming emissions per mile, while heavy-duty trucks release about four pounds per mile (ANL 2008).

The typical medium-duty truck also travels about 11,500 miles annually, so it is responsible for more than 15 metric tons of global warming emissions each year. With more annual mileage (about 36,000 miles) and lower fuel economy, the average heavy-duty vehicle is responsible for more than 65 metric tons of carbon dioxide and other heat-trapping emissions each year. And cross-country 18-wheelers put on 130,000 miles annually, so they average more than 240 metric tons of global warming emissions each year.

6.1.1. Potential and Costs of Vehicle Technologies

6.1.1.1. Efficient Choices in the Showrooms

For decades the automotive industry has been developing technologies that can safely and economically help consumers get more miles to the gallon while driving cars, minivans, pickups, and SUVs of all shapes and sizes. These off-the-shelf technologies include turbocharged direct-injection gasoline engines, high-efficiency automatic-manual transmissions, engines that shut off instead of wasting fuel while idling, improved aerodynamics, and better tires (among many others). More advanced vehicles, such as hybrids, can push fuel economy even further (see Figure 6.2).

These technologies will deliver vehicles with the same safety, utility, and performance consumers enjoy today (Gordon et al. 2007; Friedman, Nash, and Ditlow 2003). Other technologies, such as high-strength steel and aluminum and unibody construction, can boost fuel economy while actually making highways safer (Gordon et al. 2007; Friedman, Nash, and Ditlow 2003).

Automakers are already including many of these technologies on vehicles individually (see Appendix F online for examples). But not until they are packaged together can they deliver a substantial boost in fuel economy.

A variety of studies have looked at this potential and the associated costs (Kliesch 2008). For example, a

FIGURE 6.2. Fuel Economy Potential for Cars, Minivans, SUVs, and Pickups

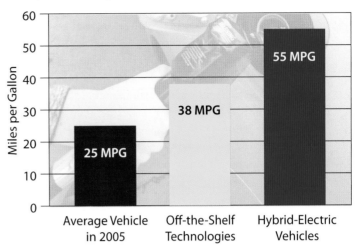

While the average vehicle in 2005 reached only 25 miles per gallon on government tests, tapping into off-the-shelf technologies already in the hands of automakers could boost the fuel economy of our cars and light trucks to 38 mpg. Adding hybrid electric vehicle technology on top of that could bring our fleet to more than 50 mpg. These values all assume a mix of 54 percent cars and 46 percent light trucks, which may change over time.

The average 18-wheel tractor trailer, or "big rig," is responsible for 65 metric tons of carbon emissions each year. Existing and emerging technologies can boost truck fuel economy by nearly 60 percent and significantly reduce emissions. To help guide this process, the Department of Energy developed the 21st Century Truck Partnership, a collaboration between government agencies and manufacturers of heavy-duty engines, trucks, buses, and hybrid powertrains.

2002 study from the National Academies of Science pointed to a potential for passenger vehicles to average about 37 mpg, at a cost of about $3,000. However, this study did not include the potential benefits of using high-strength steel to reduce the weight of vehicles (NRC 2002).

A more recent MIT study points to the potential for these conventional technologies to deliver a fleet of new cars and trucks that achieves 42–48 mpg while offering today's size and acceleration. These vehicles would cost about $2,200–$2,950 more than today's average vehicle (Bandivadekar et al. 2008).

A recent UCS study showed that a similar package of technologies could bring the fleet of new cars and trucks to about 38 mpg, at an extra cost of $1,700 (Kliesch 2008). That report found that even at fuel prices of just $2.50 per gallon—a conservative estimate, given that the *2008 Annual Energy Outlook* predicts notably higher fuel prices—owners would save almost $5,100 on gasoline over the average vehicle's lifetime, for a net savings of almost $3,400.[46] In other words, consumers would save thousands of

---

46 That figure assumes that a vehicle's real-world fuel economy is 20 percent lower than achieved on federal tests. It also assumes that the owner drives the vehicle 15,600 miles during the first year, declining by 4.5 percent per year (as an approximation of data from CTA 2008d on the decline in vehicle mileage with age), and a 7 percent discount rate. For more on modeled cost assumptions, see Table 6.1 and Appendix E online.

dollars while cutting carbon emissions by more than one-third.

Advanced technologies such as hybrid-electric drive trains hold even greater promise. In such a hybrid, an electric motor and a battery pack work together to provide supplemental power to the vehicle, which allows it to use a smaller engine that operates more efficiently. The electric motor/battery combination also allows the engine to shut off at stoplights, rather than wasting fuel while idling. Hybrids also employ "regenerative braking," which recovers some of the energy normally lost in braking and feeds it back into the battery. These technologies work together to improve fuel economy while maintaining vehicle performance.

If 25 percent of cars and light trucks were hybrids in 2020, with the remainder using the best conventional fuel economy technologies, the fleet could average 42 mpg.

The MIT study says that such hybrids have the potential to reach more than 70 mpg for about $5,100 more than the cost of today's conventional vehicles. The 2008 UCS analysis points to a more modest 55 mpg for advanced hybrids, at a cost of about $4,400 more than today's conventional vehicles. At a conservative $2.50 per gallon, owners of these vehicles would therefore save nearly $8,100 on fuel, or about $3,700 above the cost of the technology—while cutting carbon emissions in half.

6.1.1.2. Improving Air Conditioning to Cut Heat-Trapping Emissions

While fuel combustion in vehicles produces a host of heat-trapping emissions—including carbon dioxide, nitrous oxide (N_2O), unburned hydrocarbons (HC), and methane (CH_4)— burning fuel is not the only way cars and light trucks produce heat-trapping emissions. The air conditioning systems they use to keep drivers cool leak refrigerant and require extra power.

While the amount of refrigerant that vehicles leak is small compared with the amount of carbon they emit, the heat-trapping impact of today's refrigerants is more than 1,400 times that of carbon dioxide. Replacing those refrigerants with alternatives such as

Hybrids combine an electric motor and battery pack with a combustion engine to improve fuel economy, cut carbon emissions, and save drivers money at the pump. These savings are achieved through features including "idle-off" technology, which shuts off the engine when idling so that no fuel is wasted; power assist, in which the motor boosts acceleration to allow for a smaller gasoline engine; and regenerative braking, which captures some of the energy normally lost when braking and uses it to recharge the battery.

HFC-152a, which is only 120 times as powerful as carbon dioxide in trapping heat—and taking steps to reduce leaks and improve the efficiency of the air conditioning system—can cut global warming emissions by about 8 grams per mile. And the newer refrigerant system would cost just $50 per vehicle (Hill 2003).

Scientists are continuing to develop new refrigerants. The refrigerant HFO1234yf, for example, has a remarkably low heat-trapping potential of just four times that of carbon dioxide. And that refrigerant still allows compressors in air conditioners to operate efficiently (SAE 2008).

6.1.1.3. Boosting the Efficiency of Medium-Duty and Heavy-Duty Vehicles

Long-distance, heavy-duty tractor-trailers are the largest consumers of fuel in the truck category. According

TABLE 6.1. Fuel Economy Potential and Costs Used in the Climate 2030 Blueprint

	Cars and Light-Duty Trucks	Medium-Duty Trucks	Heavy-Duty Trucks
2005 Baseline Fuel Economy (mi/gallon gasoline eq)	26	8.6	6
2020 Fuel Economy for New Vehicles (mi/gallon gasoline eq)	42	11	8
2020 Incremental Cost vs. 2005 (2006 dollars)	$2,900	$6,000	$15,800
2030 Fuel Economy for New Vehicles (mi/gallon gasoline eq)	55	16	9.5
2030 Incremental Cost vs. 2005 (2006 dollars)	$5,200	$14,900	$40,500

Notes: These potentials and costs are based on assumptions in the AEO 2008 NEMS high technology case, as modified by the authors, and modeling runs of UCS-NEMS. The values in our Blueprint case model runs may not match these levels because of limitations in the model. See Appendix E online for details.

to the U.S. Census Bureau's Vehicle and In-Use Survey Microdata, the average new tractor-trailer travels 130,000 miles per year while consuming more than 20,000 gallons of diesel (U.S. Census Bureau 2002).

Using technology available today, owners can improve the fuel efficiency of these trucks more than 10 percent

> **Medium-duty vehicles, such as those used to deliver packages, could average 16 mpg by 2030, while heavy-duty trucks could average 9.5 mpg.**

by purchasing equipment to make trailers and tractors more aerodynamic, and by choosing fuel-efficient tires. The resulting savings in fuel costs—after accounting for the cost of the upgrades—can top $20,000 over the life of a truck (Anair 2008). However, despite these savings, the trucking industry has been slow to adopt many of these technologies.

What's more, studies from Argonne National Laboratory and the Department of Energy (DOE) show that better engines and transmissions—plus advanced aerodynamics, better tires, and weight reduction—could improve the fuel efficiency of heavy-duty trucks by about 60 percent, at a cost of about $40,000 (Cooper et al. forthcoming; Vyas, Saricks, and Stodolsky 2003). That would raise the gasoline-equivalent fuel

Hydraulic hybrid vehicle (HHV) technology combines a highly-efficient diesel engine with a hydraulic propulsion system, which stores energy more efficiently than a battery. This advanced technology can improve fuel economy up to 50 percent compared with conventional diesel engines, while reducing carbon emissions by as much as one-third. UPS was the first company in the package delivery industry to purchase HHVs, and currently has seven in its fleet.

economy of the average heavy-duty truck from six miles per gallon today to more than 9.5 mpg—and reduce global warming emissions per truck by more than 36 percent (see Table 6.1).

For a typical medium-duty truck, hybridization alone—by adding a battery and an electric motor or a hydraulic motor and storage system—could improve fuel economy 40 percent.[47] That improvement could

---

47 Performance of these vehicles varies based on how and where they are used, with some estimates showing a 100 percent improvement in fuel economy for use in urban stop-and-go driving (An et al. 2000). Under typical driving conditions, analysts at Argonne National Laboratory estimate that hybrids would improve the fuel economy of such vehicles 40–71 percent (Vyas, Saricks, and Stodolsky 2003).

save 1,000 gallons of fuel per vehicle each year, with the more advanced systems paying for themselves in as little as four years.[48]

By 2030, the average medium-duty truck with a combination of conventional and hybrid technologies could raise its fuel economy by more than 80 percent, at a cost of about $15,000.[49] That would boost the gasoline-equivalent fuel economy from an average of 8.6 mpg to about 16 mpg, and reduce carbon emissions per truck by 44 percent.

6.1.1.4. Trains, Ships, and Planes

Trucks are not the only mode of freight transport that can benefit from improvements in efficiency. Better vehicle and engine technology for rail, ship, and air—along with more efficient use of these modes—can also deliver cuts in emissions between now and 2030. Trains and ships can take advantage of engine improvements similar to those for heavy-duty trucks, and can also reduce engine idling.[50] Improvements aimed at maximizing loads and optimizing routes can deliver even more gains (Stodolsky 2002).

Passenger aircraft can save fuel by using lighter-weight materials and improved engine technology. And efforts to ensure that planes fly full can provide immediate benefits from existing aircraft. These incremental improvements can boost efficiency in rail, ship, and air by 10–15 percent. Shifting freight from one mode to another and tapping alternative fuels (see below) can also reduce emissions from freight transportation.

6.1.2. Key Challenges for New Vehicle Technologies

Challenges

All the technologies and other options for travel and shipping in this chapter point to a 2030 where consumers and companies can do their part to cut heat-trapping emissions while also reducing America's oil addiction and saving money. But if past is prologue, the needed changes simply won't happen on their own. Each of the technologies and other options faces barriers to becoming mainstream.

One barrier has been low gas prices. Between 1990 and 2003, annual average gasoline prices ranged from about $1 to $1.50 per gallon (EIA 2009c). The average vehicle achieved only about 25 mpg on government tests, as automakers devoted new technology to boosting vehicle size and power instead of fuel economy (OTAQ 2008).

Gasoline would have to rise to $5–$10 per gallon just to encourage consumers to purchase vehicles that reach about 40 mpg, according to estimates of people's responses to price.[51] And the upper end of that range may be the most realistic. A recent study indicates that consumers are becoming less responsive to gas prices as household incomes rise (Small and Van Dender 2006).

A second, related, barrier is that consumers appear to be averse to risk. Purchasing a vehicle with higher fuel economy means making an investment today that will yield benefits in fuel savings over time. Yet consumers lack information on future fuel prices, and are unsure about how long they will own the vehicle and other factors—and that creates uncertainty about whether their investment in fuel economy will pay off. Recent research indicates that the result is a market failure: consumers choose fuel economy lower than what makes sense given the costs and benefits (Greene, German, and Delucchi 2009). That is, consumers are sensitive to sticker prices, placing a greater emphasis on up-front costs despite potential longer-term benefits.

A final barrier has been a lack of options for consumers. They have historically had to shift to smaller or less powerful vehicles if they wanted much better fuel economy. Only with the recent introduction of hybrids have consumers been able to choose a vehicle with better fuel economy that also has the size and acceleration of the vehicle they already own.

Other automakers were caught off guard by the success of the Prius when Toyota first introduced it in

---

48 This figure is based on 30,000 annual miles driven, a 40 percent improvement in fuel economy, an incremental cost of $10,000, a fuel price of $2.50 per gallon, and a 7 percent discount rate.

49 This figure is based on improvements in conventional and hybrid technologies described in Vyas, Saricks, and Stodolsky 2003. Even greater improvements may be available, as indicated by the goal of the DOE's 21st Century Truck program: to improve the fuel efficiency of medium-duty (Class 6) trucks by a factor of three (DOE 2000).

50 Hybrid tugboats and switcher locomotives designed to reduce or eliminate idling are two examples of where these technologies are already gaining traction.

51 This figure assumes that the long-run price elasticity of demand for lower fuel intensity (fewer gallons per mile) ranges from 0.2 to 0.4 (based on Brons et al. 2006; Small and Van Dender 2006; Espey 2004; Goodwin, Dargay, and Hanly 2004; Dahl 1993). The figure also assumes fixed household income. Note that consumers would purchase fewer vehicles and drive less because of the higher prices.

Japan in 1997 and in the United States in 2000, when gas prices were still quite low (Sperling and Gordon 2009). It is difficult for consumers to show demand for a product with high fuel economy if it is not on the market.[52]

6.1.3. Key Policies for Putting Better Vehicle Technology to Work

Policies

If the main barrier to better fuel economy and less global warming emissions is simply low gas prices, then the policy solution could be to raise those prices. But as the previous section showed, gas prices might have to rise to $10 per gallon to encourage consumers to move just to 40 mpg vehicles. Even higher gas prices would be needed to spur a wider move to the better fuel economy offered by hybrids. That suggests another barrier to the use of technologies to improve fuel economy: the political difficulty of creating policies that will raise gas prices enough to deliver the benefits of those technologies.

That political barrier does not mean we should abandon attempts to create policies that influence gas prices. Instead, it means that we cannot rely on them on exclusively. Indeed, accurately pricing gasoline is essential to capture the costs of smog, global warming, the U.S. military presence in the Middle East, and other externalities that gas prices do not now reflect.

An economywide cap-and-trade policy that includes the Blueprint's complementary policies for energy and transportation will gradually add up to $0.55 per gallon to the price of gasoline. Such a policy will provide modest cuts in carbon emissions by spurring people to reduce the number of miles they drive and encouraging automakers to increase fuel economy somewhat. Such a policy will also help internalize the costs of the impact of gasoline consumption on our climate. However, that policy alone will not deliver the full potential benefits of better technologies for both conventional and hybrid vehicles.[53]

Instead, policies that require or reward better vehicle performance—whether higher fuel economy or lower carbon emissions—can deliver on this potential and overcome all three barriers. As Greene, German, and Delucchi (2009) note, such policies remove or reduce the uncertainty associated with the benefits of more

Over the past 30 years, Corporate Average Fuel Economy (CAFE) standards have made America's cars and trucks go farther on a gallon of gas, saving consumers hundreds of billions of dollars in gas costs. Under a proposal announced by President Obama in May 2009, CAFE standards would be coordinated with the nation's first system to regulate carbon emissions from cars and light trucks—helping to further reduce global warming emissions, consumer costs, and America's oil dependence.

efficient vehicles. Such policies also guarantee that consumers will be able to choose higher-fuel-economy or lower-polluting vehicles in all types and sizes, not just small cars.

The federal government has relied on performance-based standards for about 40 years to reduce smog-forming and toxic pollution from cars and trucks (such as grams per mile for Tier 1 and Tier 2 vehicles). Such standards have cut these pollutants by well over 90 percent compared with emissions from cars and trucks in the 1960s. Federal performance standards have also cut gasoline use in cars and trucks.

The federal government created the corporate average fuel economy (CAFE) standards more than 30 years ago, to boost the fuel economy of cars and trucks in response to an oil embargo. If car companies do not meet those standards, they are subject to a fine.[54] Had these standards not been around, and had consumers been stuck with the same fuel economy choices from the 1970s—when vehicles averaged about 15 mpg on government tests, versus about 25 mpg in 2007—they would have needed to purchase another 40 billion

---

52 This resembles the experience with airbags. Automakers fought a federal requirement that they install airbags, but now compete based on the number onboard in response to consumer interest in safety.

53 This analysis assumes base gas prices of $2.50–$3.50 per gallon. Adding even $0.55 per gallon would not raise the price enough to exhaust conventional technology, let alone hybrids, based on elasticity values for gasoline demand of 0.2–0.4.

54 The fine is $5.50 per 0.1 mpg below the standard, multiplied by the manufacturer's sales volume (40 CFR 32912).

gallons of gasoline in 2007, at a cost of more than $100 billion.[55]

While CAFE standards are clearly saving consumers money today, they have been nearly stagnant for the past 20 years. That changed at the end of 2007, when Congress required that America's cars and trucks average at least 35 mpg by 2020.[56]

BOX 6.1.

The Advantages of Regulating Vehicle Emissions versus Fuel Economy

Today the National Highway Traffic Safety Administration (NHTSA) regulates the fuel economy of vehicles through its CAFE (Corporate Average Fuel Economy) standards, but those do not directly cap heat-trapping emissions from vehicles. Having the Environmental Protection Agency (EPA) regulate carbon emissions directly would offer several advantages:

- The EPA can set long-term standards, while NHTSA can set standards for only five years at a time. This limits NHTSA to technologies that are available today, and robs automakers of the regulatory certainty that would help them direct their long-term investments.
- The EPA can consider the potential of all technologies to reduce the emissions of vehicles and fuels, while the law forbids NHTSA from accounting for the impact of alternative fuels on fuel economy standards.
- An EPA-based standard for global warming emissions would guarantee a shift away from oil, one of the most carbon-intensive fuels on the market. NHTSA's fuel economy standards, in contrast, do not guarantee cuts in carbon emissions, because the agency must use complicated formulas that reward displacement of oil alone. For example, NHTSA assumes that natural gas will reduce the use of petroleum-based fuels by more than 80 percent. However, such a substitution would reduce carbon emissions by only about 15 percent. NHTSA's process also overstates the impact on oil consumption and carbon emissions of a shift to diesel by at least 10 percent.
- The EPA would regulate global warming gases beyond CO_2, such as the refrigerants used in vehicle air conditioning systems, as well as nitrous oxide and methane. NHTSA's regulations ignore those emissions.

An alternative to CAFE standards is to target heat-trapping emissions from cars and trucks directly. Such standards can tap a broader set of solutions than fuel economy, including better air conditioning systems and fuels, while also saving consumers money and cutting the use of gasoline.

California and 14 other states have already adopted standards requiring new cars and trucks to cut global warming emissions by about 30 percent in 2016, and California is considering stronger standards for 2020.[57] These state standards represent a more aggressive attempt to address global warming emissions, but they still fall short of the potential for the technologies outlined earlier in this chapter. As of 2009, the U.S. Environmental Protection Agency (EPA) was also considering adopting standards on global warming emissions for cars and trucks nationwide (see Box 6.1).

If we are to reap the benefits of technologies that are available now—including low-carbon fuel (see below)—national standards would have to require new cars and light trucks to average no more than 200 grams of global warming emissions per mile in 2020, and no more than 140 grams per mile in 2030 (see Table 6.2).[58]

The 2020 level is based on enabling conventional passenger vehicles to achieve 38 mpg, and spurring hybrids to account for 25 percent of the car and light-truck market. The combination would produce an average fuel economy for passenger vehicles of 42 mpg. The 2020 level also assumes that all automakers would install better air conditioning, which would cut 8 grams per mile by 2015, and that the federal government would create a 3.5 percent low-carbon fuel standard (see below). The 2020 standard also accounts for the fact that today's vehicles emit about 1.9 grams per mile of heat-trapping gases other than carbon dioxide.

The 2030 standard is based on near-complete market penetration of hybrids, 20 percent penetration of plug-ins (see below), full adoption of air conditioning improvements, a 10 percent low-carbon fuel standard, and the same 1.9 grams per mile in other heat-trapping emissions.[59]

States and the EPA can establish similar standards to reduce emissions from medium- and heavy-duty vehicles. Those standards would have to account for the

---

55 This figure is based on 2.67 trillion miles traveled in cars and trucks in 2007, gasoline at an average of $2.843 per gallon (EIA 2009d), a rebound effect of 10 percent, and estimated on-road fuel economies of 13.1 mpg (OTAQ 2008) versus 20.2 mpg (EIA 2008a).

TABLE 6.2. Standards for Vehicle Global Warming Emissions

	Cars and Light-Duty Trucks	Medium-Duty Trucks	Heavy-Duty Trucks
2005 Baseline Global Warming Emissions (g/mi CO$_2$ eq)[a]	**372**	**1,038**	**1,489**
Fuel Economy (mi/gallon gasoline eq)	24	8.6	6
Non-CO$_2$ Emissions Estimate (g/mi CO$_2$eq)	2	5	8
2020 Standard for Global Warming Emissions (g/mi CO$_2$eq)[a]	**198**	**777**	**1,072**
Fuel Economy (mi/gallon gasoline eq)	42	11	8
CO$_2$ Emissions with Current Gasoline (g/mi CO$_2$eq)[b]	212	808	1,111
Non-CO$_2$ Emissions Estimate (g/mi CO$_2$eq)[c]	2	5	8
Credit for Improved A/C (g/mi CO$_2$ eq)[d]	-8	-8	-8
Credit for Low-Carbon Fuel Standard (g/mi CO$_2$eq)[e]	-7	-28	-39
2030 Standard for Global Warming Emissions (g/mi CO$_2$eq)[a]	**139**	**497**	**842**
Fuel Economy (mi/gallon gasoline eq)	55	16	9.5
CO$_2$ Emissions with Current Gasoline (g/mi CO$_2$eq)[b]	162	555	935
Non-CO$_2$ Emissions Estimate (g/mi CO$_2$eq)[c]	2	5	8
Credit for Improved A/C (g/mi CO$_2$ eq)[d]	-8	-8	-8
Credit for Low-Carbon Fuel Standard (g/mi CO$_2$eq)[e]	-16	-56	-94

Note: Values may not sum properly because of rounding.

a We calculated global warming emissions as the sum of CO$_2$ and non-CO$_2$ emissions from today's gasoline, minus cuts in emissions from the use of better air conditioning and low-carbon fuels.

b In converting fuel economy into CO$_2$ equivalent, we assumed 8,887 grams of CO$_2$ per gallon of today's gasoline burned.

c We scaled up estimates of non-CO$_2$ heat-trapping emissions for medium- and heavy-duty trucks from those for light-duty vehicles based on relative fuel consumption. We expect to update these numbers as more accurate data become available. These estimates do not include black carbon.

d Note that 8 grams per mile is a conservative estimate for cars and light trucks based on Bedsworth 2004 and CARB 2008. We have no data for medium- and heavy-duty vehicles. However, given that they have larger air conditioning systems (and thus greater potential for absolute savings) but travel farther (reducing the per-mile benefit), we used 8 grams per mile as a rough value pending more information.

e All fuels achieve the average low-carbon standard in Table 6.4.

---

56 As required by the 2007 Energy Independence and Security Act, online at *http://frwebgate.access.gpo.gov/cgi-bin/getdoc. cgi?dbname=110_cong_bills&docid=f:h6enr.txt.pdf*. Our Reference case includes this policy, which delivers significant cuts in carbon emissions. Reductions in emissions from the transportation sector under the Blueprint are therefore notably lower than they would otherwise be.

57 Under the Clean Air Act, California can adopt its own vehicle standards. Other states must choose between California's standards or those of the federal government. However, the latter has yet to regulate global warming emissions from vehicles. California's proposed stronger standards would reduce heat-trapping emissions from vehicles by more than 40 percent, according to estimates. For more information, see Tables 4 and 6 at *http://www.climatechange.ca.gov/publications/arb/ARB-1000-2008-012/ARB-1000-2008-012.PDF*.

58 Under this approach, automakers would receive credit for selling vehicles that use ethanol, hydrogen, or electricity, if they reduce global warming emissions from the production and delivery of the fuel.

59 The factors used for air conditioning and non-CO$_2$ emissions for the 2020 and 2030 standards are consistent with those used by the California Air Resources Board (CARB 2008), and are conservative in the case of air conditioning, given that newer refrigerants with very low potential for trapping heat are likely to enter the market soon.

BOX 6.2.

Promising Policies the Blueprint Case Did Not Include

Feebates and the California Clean Car Discount. Besides vehicle performance standards, a system of financial carrots and sticks can encourage automakers to make better vehicles, and consumers to purchase them. Feebates create surcharges on vehicles with more heat-trapping emissions, and use the proceeds to pay for rebates on cleaner vehicles. California is considering adopting such a system. A 2007 study suggests this approach could encourage greater use of technology to cut carbon emissions while saving consumers money (McManus 2007).

Air, rail, marine, and off-road standards. Limits on global warming emissions can also apply to planes, trains, ships, and off-road vehicles—all of which can benefit from technologies to improve efficiency, including hybridization in some cases, and can tap into cleaner fuels and improved air conditioning systems.

The aviation sector is responsible for 10 percent of heat-trapping emissions from transportation, so regulators should not ignore it. And construction and other off-road equipment present an important opportunity for further emissions cuts, as that equipment uses about 7 percent of all diesel fuel.

Without such policies, however, baseline improvements are still likely. The efficiency of rail transport (ton miles per Btu) is likely to rise by a modest 10 percent between 2005 and 2030, marine transport (ton miles per Btu) by 12 percent, and passenger aircraft (passenger miles per gallon) by 16 percent.

Lower speed limits. When cars and trucks travel at highway speeds, they use a lot of fuel to overcome aerodynamic drag. Lowering maximum speeds can save fuel and reduce carbon emissions.

For example, reducing vehicle speeds from 70 to 65 mph could cut highway fuel use and carbon emissions by 8–10 percent per mile (CTA 2008). A forthcoming study by the North East States Center for a Clean Air Future similarly finds that lowering the maximum speed of a tractor-trailer on a typical long-haul trip to 60 miles per hour could reduce carbon emissions by 4 percent. Such a vehicle can require 5–10 percent more fuel traveling at 70 mph than at 65 (DOE 2000).

Freight transport standards. California is considering a suite of regulatory and voluntary measures aimed at reducing carbon emissions from the use of planes, trains, and ships to transport freight. Those measures could cut emissions from the movement of goods by 3.5 million metric tons of carbon dioxide equivalent by 2020—a 20 percent reduction.

Efforts that could help meet that goal include reducing the speeds of oceangoing vessels, connecting docked ships to the electricity grid, using better hull and propeller maintenance practices, reducing idling of cargo-handling equipment and relying on electricity to power it, setting energy efficiency standards for refrigerated trailers and containers, and using zero-emission rail technologies (CARB 2008a).

---

many different uses of such vehicles, as well as the numerous manufacturers of engines, truck chassis, and trailers (Lowell and Balon 2009).[60]

Today's medium-duty vehicles produce more than 1 kilogram of global warming emissions per mile, not counting upstream emissions. Based on the vehicle technologies noted above and the potential for cleaner fuels, state and federal standards should allow such vehicles to release no more than 780 grams of heat-trapping emissions per mile in 2020, and no more than 500 grams per mile in 2030—representing cuts of 25 percent and 50 percent, respectively.

For heavy-duty vehicles, standards should be no more than 1,075 grams per mile in 2020, and no more than 840 grams per mile in 2030. Those standards would cut carbon emissions more than 25 percent and 40 percent, compared with today's emissions of about 1,500 grams per mile.

These standards assume full adoption of available technology for improving fuel economy by 2030, and

---

60 The EPA can regulate carbon emissions for medium- and heavy-duty vehicles under the Clean Air Act. The Energy Independence and Security Act of 2007 authorized the National Highway Traffic Safety Administration to establish fuel economy standards for medium- and heavy-duty vehicles, based on a National Academy of Sciences study now under way. Japan established fuel efficiency standards for heavy-duty vehicles in 2005.

3.5 percent and 10 percent low-carbon fuel standards for 2020 and 2030. The standards also assume that better air conditioning systems in medium- and heavy-duty trucks would reduce emissions 8 grams per mile, though we lack data on actual values. (The air conditioning systems in these vehicles are larger, so their emissions are higher than those from cars and light trucks. However, medium- and heavy-duty vehicles also travel more miles per year, so we assume that the per-mile emissions from all the vehicles are about the same.)

6.2. Smart Fill-Ups: Switching to Low-Carbon Fuel

Americans use the equivalent of nearly 220 billion gallons of gasoline every year (EIA 2008a). Cars and light trucks use the largest portion of that amount: 62 percent, or 140 billion gallons. Medium-duty and heavy-duty trucks are next, at 18 percent, followed by airplanes at 10 percent. Rail, shipping, military, and other uses account for the last 10 percent (see Figure 6.3).

Even if the nation takes aggressive measures to increase the efficiency of vehicles and reduce the number of miles we travel, we will still need the equivalent of about 200 billion gallons of gasoline by 2030 as the population and economy continue to grow. If transportation is to do its part in cutting U.S. carbon emissions close to 60 percent by 2030, we cannot continue to fill up almost exclusively on petroleum products as we do today.

To make these deep cuts while continuing to strengthen our economy, we must tap into transportation fuels that do not release significant amounts of carbon. Biofuels (fuels produced from plants), electricity, and hydrogen all have the potential—if produced in a sustainable manner and without significant impacts on land use—to both cut carbon emissions from transportation and curb our country's dependence on oil.

 **6.2.1. Potential and Costs of Low-Carbon Fuel**

6.2.1.1. Biofuels for Today and Tomorrow

You probably already use biofuels when you drive your car. To cut down on smog or substitute for gasoline, fuel makers now blend ethanol—an alcohol—into

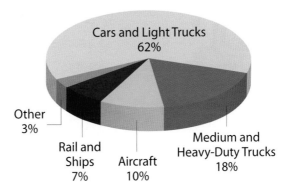

FIGURE 6.3. Petroleum Use in Transportation (2005)

218 billion gallons gasoline equivalent

Cars and Light Trucks 62%
Other 3%
Rail and Ships 7%
Aircraft 10%
Medium and Heavy-Duty Trucks 18%

Source: EIA 2008.

gasoline, where it accounts for 4–5 percent of what Americans buy (EIA 2008a).[61]

As noted, biofuels are made from plants: ethanol is made from corn or sugar cane, for example, and biodiesel from vegetable oil. Existing vehicles can use small amounts of ethanol and biodiesel, but must be modified to use larger amounts.[62] However, enabling cars to use up to 85 percent ethanol (E85) adds only $50–$100 to their price (DOT, DOE, and EPA 2002).

New technologies nearing commercialization will create fuel based on other types of plant material, such as wood, grass, and waste products from agriculture, forestry, and landfills. Some of these technologies can be used to make biofuels that are even more compatible with gasoline and diesel engines, potentially requiring no engine modifications at all.

Because the carbon in biofuels comes from plants rather than fossil fuels, they could theoretically provide carbon-free transportation—if regrown crops absorbed the CO_2 emitted from tailpipes. However, the true picture is not that simple. To get a complete understanding of the carbon and other heat-trapping emissions from fuel, we need to look at its full life cycle, which includes all the emissions caused directly or indirectly by its production, distribution, and use.

For ethanol based on corn, that means accounting for the fertilizer and pesticide used to grow it, the energy used to convert the corn to fuel, the tractors, trains, and tanker trucks used to move the fuel around, and

---

61   Ethanol is blended at 6–7 percent to cut down on smog, but is not used everywhere year-round.

62   Ethanol blended with gasoline up to 10 percent is sold as E10, or gasohol. Biodiesel can be blended up to 5 percent under today's warranties, although some vehicles use blends of up to 20 percent.

Biofuels made from food crops, especially corn and soybeans, require a lot of energy to produce and can actually create more global warming emissions than traditional gasoline. However, cellulosic biofuels—made from sugars in the plant walls of corn cobs and stalks, grasses, wheat straw, and other plants—have the potential to cut carbon emissions by 80 percent or more compared with gasoline. This South Dakota farmer is harvesting switchgrass, a prairie grass native to the Midwest that can be grown on land not typically used for agriculture.

finally the emissions from vehicles' tailpipes. We also need to account for heat-trapping emissions from indirect effects, such as the expansion of agricultural production to produce more biofuels and the resulting changes in land use. When we add up all of these sources of emissions, today's generation of biofuels—which are made from food crops, especially corn and soybeans—offer little opportunity to reduce carbon emissions, and may even increase them (Fargione et al. 2008; Searchinger et al. 2008; UCS 2008a).

To deliver biofuels that can provide significant cuts in carbon emissions, the industry is moving to second-generation technologies. Cellulosic biofuels, for example, are produced from the materials that compose the cell walls of all plant matter. That means corn cobs and corn stalks, grasses, wheat straw, and sawdust can all be used to make fuel. A closely related technology called biomass-to-liquids (BTL) can be used to make a replacement for diesel fuel from plants, wood, or even the carbon-containing portion of municipal garbage, such as lawn clippings and used plastic.

These next-generation biofuels could cut carbon emissions from transportation by 60–80 percent or more (ANL 2008). However, these cuts rely on using waste products, grasses, or other crops grown on land that does not directly or indirectly displace food crops. These restrictions mean that the amount of low-carbon biofuels we can make from domestic resources is significantly less than some analysts estimate (Perlack et al. 2005; Greene et al. 2004).

Further, some of those resources will be used in power plants to generate cleaner electricity. Based on the domestic potential given land-use restrictions and other exclusions outlined in Chapter 2, the transportation sector could tap about 280 million tons of biomass for conversion into biofuel.[63] At a conversion rate of 110 gallons of ethanol per ton of biomass, we estimate that the ethanol equivalent of about 30 billion

---

63 This is the combination of 158 million tons of agricultural residues and 121 million tons of biofuel crops grown without inducing direct or indirect changes in land use.

gallons of low-carbon biofuels could be available in 2030—enough to displace about 20 billion gallons of gasoline (see Table 6.3).[64]

Biofuels have historically been more expensive than gasoline. However, the next generation of biofuels has the potential to be cost-competitive with gasoline prices of $2.70–$3.00 per gallon (Anden et al. 2002), and some studies point to the potential for even lower costs (ASES 2007; Greene et al. 2004).

The nation could produce about 30 billion gallons of low-carbon biofuels in 2030, displacing 20 billion gallons of gasoline.

Biofuel makers can achieve cost-competitiveness through greater efficiency, lower production costs, and "biorefinery" approaches, which combine ethanol production with the production of electricity, heat, and animal feed. Such technologies can lead to true low-carbon biofuels that can deliver dramatic cuts in carbon emissions while allowing consumers to at least break even and possibly even save money compared with future gas prices.

6.2.1.2. An Electrifying Transportation Future

There is a lot of excitement today about the potential for drivers to not only own a hybrid vehicle but to plug it in and recharge the battery pack from the electricity

The leaves and stalks left behind after the corn harvest—known as corn stover—have substantial potential as a biofuel. Increasing our use of biofuels in vehicles and power plants can help reduce our dependence on fossil fuels.

grid. Such "plug-in" hybrids would have the fuel efficiency and range of a conventional hybrid, but would rely on a larger battery pack to tap into electricity as another potentially low-carbon fuel.

The next step after plug-ins is to get rid of the engine and have an all-electric vehicle, which would require an even bigger battery pack. If based on most

TABLE 6.3. A Look at Cellulosic Ethanol in 2030

Resource, Yield, and Potential		Costs	
Biomass Resources Available for Transportation (million tons)	280	Fixed Production Costs (in 2006 dollars per gallon)	$0.128
Ethanol Yield (gallons per ton)	110	Non-Feedstock Variable Costs (in 2006 dollars per gallon)	$0.17
Maximum Biofuel Potential (billion gallons ethanol equivalent)	30	Initial Capital Cost (in 2006 dollars per gallon of capacity)	$1.99

Note: In our Blueprint analysis, actual production of cellulosic ethanol may be lower, as it competes with biomass-to-liquids technology for access to biomass resources. However, the total volume of low-carbon biofuels will be similar.

64 The 110 gallons of ethanol per ton of biomass is based on information in ASES 2007 (90 gallons per ton) and Greene et al. 2004 (100–126 gallons per ton). Ethanol and gasoline equivalents are used for convenience, and because the federal renewable fuel standard uses ethanol equivalence. The actual volume of the biofuel will vary with the density of the product.

Plug-in hybrids have real potential to reduce carbon emissions and fuel costs from passenger vehicles. A plug-in with an all-electric range of 30 miles could supply the average driver with about half of his or her driving needs using electricity alone. If that electricity comes from a battery charged from renewable electricity sources, carbon emissions will be at least 70 percent lower than today's vehicles.

battery technologies available today, such a vehicle would give up a significant amount of range. However, the electric power train in such a vehicle is more than four times as efficient as that in a conventional vehicle, and about two times as efficient as that of a hybrid.

Electric vehicles have not advanced past small-scale commercialization because of high battery costs, short lifetimes, and limited range. Advances in battery technology—such as lithium ion batteries similar to those used to run laptops—have brought battery-electric vehicles closer to commercial reality. However, plug-in hybrids are more likely to reach sales in the millions first (Kalhammer et al. 2007).

General Motors, Toyota, and Ford have all announced plans to put early-model plug-ins into small-scale production between 2010 and 2012, and companies that convert conventional hybrids into plug-ins are already making a few available (Ford Motor Company 2009; Toyota Motor Sales 2009; GM 2008). GM has noted that its first plug-in will be a car that can travel about 40 miles on the battery alone, and that it will cost around $40,000 in small volumes—about a

$20,000 premium over the cost of a conventional car (Gonzales 2008). Toyota and Ford are targeting plug-ins with smaller all-electric ranges.

Given the variety of possible plug-in configurations, and their potential for an all-electric range of 5–10 miles to 40 miles or more, the Blueprint analysis assumes that the average plug-in will have an all-electric range of about 30 miles. That would allow the average driver to satisfy about half of his or her driving needs using electricity alone (Komatsu et al. 2008; Santini and Vyas 2008; Tate, Harpster, and Savagian 2008; EPRI 2007a).

A combination of fuel cell vehicles and plug-in hybrids could account for at least 20 percent of the car and light-truck market by 2030.

Based on recent studies, we expect a typical plug-in with an all-electric range of about 30 miles to cost about $8,650 more than today's conventional vehicles, when sold at high volume—or about $4,250 more than hybrids (Bandivadekar et al. 2008; Duvall 2002). If a driver can satisfy half of his or her driving on electricity at $0.10 per kilowatt-hour (equivalent to about $3.60 per gallon of gasoline), and gasoline is $2.50 per gallon, that driver will save more than $9,500 on fuel over the vehicle's lifetime compared with today's conventional vehicles, for a net lifetime savings of $850.[65]

As with biofuels, we must count all the emissions released during the life cycle of fuels used for plug-ins. For electricity, that means going back to the power plant. As a result, cuts in emissions from such vehicles can vary significantly, depending on where the electricity comes from. Given the average mix of fuels now used to produce electricity, a good plug-in will cut carbon emissions by about 55 percent compared with today's vehicles—if half the vehicle's miles come from the battery (ANL 2008). If the battery is recharged from a 2030 grid with 70–80 percent lower carbon emissions, the vehicle's emissions would be at least 70 percent lower than today.

---

65   While electricity is more expensive in this case, driving on electricity alone is much more efficient, so costs are lower per mile. In this plug-in example, electricity costs $0.033 per mile, while gasoline costs $0.057 per mile. The analysis assumes that the plug-in uses 0.33 kilowatt-hour per mile when operating on battery electricity. The analysis also assumes that the federal test fuel economy is 55 miles per gallon when the vehicle runs as a conventional hybrid, but that on-road fuel economy is 20 percent lower. The analysis also assumes that the vehicle is driven 15,600 miles during the first year, declining at 4.5 percent per year, and a 7 percent discount rate. For more on modeled cost assumptions, see Table 6.1 and Appendix E online.

Significant uncertainties remain as to how quickly plug-ins will be ready to enter the market, and how quickly that market will grow. The first mass-market hybrid, the Prius, introduced in 1997, accounted for about 2 percent of the U.S. market in 2007, 10 years later (Ward's Auto Data n.d.). If we assume that plug-ins—which Toyota and GM expect to introduce in 2010—parallel the significant success of conventional hybrids, they could reach 2 percent of the U.S. market by 2020. From there, a 25 percent average annual growth rate would have plug-ins capturing 20 percent of the market by 2030, given the proper incentives.

6.2.1.3. Hoping for Hydrogen Transportation

The excitement expressed today about plug-ins belonged to hydrogen fuel cells about three to five years ago. Like batteries, fuel cells provide electricity, but instead of storing it directly, a fuel cell generates it from hydrogen and air. That enables fuel cell vehicles—unlike battery-electric vehicles—to have driving ranges that approach those of today's vehicles. For example, the Honda Clarify FCX fuel cell vehicle is rated as having a range of 280 miles (American Honda Motor Company 2009).

Hydrogen fuel cells lost some of their luster when the technology did not deliver as quickly as automakers had hoped (much like battery-electric vehicles in the early 1990s). Despite that, automakers have made significant progress in lowering the costs of fuel cells and increasing their durability, and they could see small-scale production as early as 2015 if given enough support (NRC 2008; Kalhammer et al. 2007).

The incremental cost of fuel cell vehicles is quite high today, but could come down significantly over time. Reports from the National Research Council and MIT indicate that the incremental cost of fuel cells produced in large volumes can be similar to the costs we have presented for plug-ins (Bandivadekar et al. 2008; NRC 2008). The MIT study shows that the carbon benefits will be similar as well.

As with plug-ins, many uncertainties remain around the future of mass-market fuel cell vehicles. The National Research Council study suggests that sales could account for slightly more than 20 percent of the market by 2030.

6.2.1.4. Cleaner Gasoline and Diesel

While alternatives to petroleum fuels have clear potential, we will undoubtedly be using gasoline and diesel for decades to come. Given that, the nation also needs

to reduce the emissions associated with those fuels by improving the efficiency of refineries.

In 2000, the petroleum industry created a technological vision that pointed to the potential for a 10 percent improvement in refinery efficiency by 2020 (API 2000). A 2005 study from Lawrence Berkeley National Laboratory suggested paths to improve the efficiency of refineries across the country by 10–20 percent that would also cut costs (Worrel and Galitsky 2005). Based on today's efficiency levels for refineries of about 90 percent (Wang 2008), a 10–20 percent improvement in efficiency would lead to a 1–2 percent reduction in carbon emissions from gasoline.

6.2.1.5. Avoiding Dirty Fuels

While new low-carbon alternatives can reduce global warming emissions from fuel, new high-carbon fuels can easily wipe out these hard-won gains. Development of oil shale, tar sands, heavy crudes, and coal-to-liquid (CTL) fuels can all substantially increase the upstream emissions associated with gasoline and diesel fuels.

The life-cycle carbon emissions from liquid transportation fuel made from coal are double those of conventional petroleum (Bartis et al. 2008). In fact, displacing a gallon of petroleum fuel with a gallon of CTL more than cancels out the benefits of displacing a gallon of gasoline with low-carbon biofuel or electricity.

Emissions from crude oil recovered from tar sands and oil shale are also much higher than those from

Oil and gas refineries could become at least 10 percent more efficient by 2030, according to Lawrence Berkeley National Laboratory.

conventional crude—and would mean that we would backslide on our path to cleaner alternatives to oil. Projections of these sources producing more than 6 million barrels a day by 2035 (Task Force on Strategic Unconventional Fuels 2006) suggest that avoiding these dirty fuels is just as important as developing better alternatives, to ensure steady progress in cleaning up our fuel supply.

6.2.1.6. All of the Above

The reality is that no one can predict which of the lower-carbon fuels will win out. In fact, the most likely outcome is a mix of different options. Electricity and hydrogen are not well-suited for use in planes, ships, or big trucks, but will work well for cars and light trucks, and for some medium-duty trucks that operate mostly on city streets. Biofuels will work well in all parts of the transportation sector, but are a more limited resource if they are to remain truly low-carbon, so we might best use them where neither electricity or hydrogen work best (such as in airplanes).

Further complicating matters, the next 10 years will probably see some trial and error, wherein every sector tries all the low-carbon options. Based on that competition, plug-ins, fuel cell vehicles, and battery-electric vehicles could all emerge successful. If so, these electric-drive vehicles could account for one-third or even more of the market by 2030.[66]

66 California has a different mix of vehicles and sources of electricity. However, an analysis by the California Energy Commission shows that plug-in, battery, and fuel cell vehicles together could capture about 33 percent of the state's market in 2030 (Bemis 2008).

BOX 6.3.

SUCCESS STORY
Jump-Starting Tomorrow's Biofuels

The sustainable harvesting of perennial grasses and forestry and agricultural waste products to make biofuels can be an important part of our low-carbon future.

Corn and soybeans, grasses and wood chips, even municipal waste dumps—what do they have in common? In a world seeking to trim its dependence on the fossil fuels that, when burned, overload our atmosphere with carbon, these items all have the potential to be turned into vehicle fuels. Unfortunately, biofuels are not all created equal, at least not when it comes to curbing carbon emissions. The future of biofuels depends on making the right choices today.

The basic technology to extract liquid fuel from plants like wood and grass and other forms of biomass has existed for decades, but has not been cost-effective compared with the cost of producing gasoline and diesel. Driven primarily by an influx of federal dollars,

production of corn ethanol has grown to 3 percent of the fuel used in U.S. passenger cars and trucks (EIA 2008a). However, land, water, and other resource constraints limit the potential of food-based biofuels—such as corn ethanol and soy biodiesel—to reduce the carbon footprint of our transportation fuels (UCS 2008a; UCS 2008b). A brighter future for biofuels requires technologies for making fuel from wood chips, grasses, and waste products—and then developing sustainable sources of these feedstocks.

Recent breakthroughs in biological research, combined with government support, are bringing us closer to making fuel from plant leaves, stems, and stalks (cellulosic biofuels) a commercial reality. Several new companies are making the transition from laboratory testing to pilot manufacturing plants.

Mascoma, for example, has built a pilot plant in Rome, NY, that can make half a million gallons of biofuel a year from wood chips. Verenium has opened a 1.4-million-gallon-a-year plant in Jennings, LA, to make ethanol from crushed sugar cane stalks. Both these plants use biochemical processes to break down cellulose into ethanol (Verenium 2009; LaMonica 2008). Bluefire Ethanol in Southern California is using a different approach—breaking down cellulose in municipal waste to make sugar via acid hydrolysis—and will begin construction this year of a 3.7-million-gallon-a-year facility in Lancaster, CA (Bluefire Ethanol 2008).

However, while exciting, these pilot plants are far too small to meet the nation's demand for cellulosic biofuels. In comparison, corn ethanol facilities often produce 100 million gallons a year or more, and petroleum refineries

On the other hand, given the fact that putting the maximum feasible number of fuel cell vehicles on the road over the next decade will cost about $50 billion, success for all these options may be difficult. Still, together they should be able to account for 20 percent of the car and light-truck market, as the National Research Council noted was possible for fuel cell vehicles alone (NRC 2008).

To supply the energy that transportation needs, the nation will have to tap renewable electricity, clean hydrogen, and low-carbon biofuels while avoiding fuels

can be 20 times that size (EIA 2008g; RFA 2009). The next step is commercial-scale facilities for cellulosic ethanol.

Range Fuels in Soperton, GA, is the top contender in the race to produce such fuel at a scale of tens of millions of gallons a year. The company has broken ground on a facility, and expects to begin using high-temperature gasification to turn the cellulose in waste wood chips into liquid fuel in 2010. Range Fuels plans an initial capacity of 20 million gallons a year, eventually expanding to 100 million gallons a year (Range Fuels 2007).

A competing approach to large-scale production of cellulosic ethanol relies on microorganisms to break down the cellulose. Using this technology, Mascoma's facility in Kinross, MI, is scheduled to produce 20 million gallons a year of ethanol from wood waste by 2011 (Reidy 2008). And Verenium plans to build a commercial-scale cellulosic ethanol facility in Highlands County, FL, to convert grasses into perhaps 36 million gallons of ethanol a year.

The variety of technologies, feedstocks, and locations tapped by these promising projects improves the chances that one or more will produce the breakthroughs that move the approach from laboratory to market. Scaling up next-generation biofuels from less than a million gallons a year in 2008 to more than a billion is essential if biofuels are to be players in America's low-carbon future.

from tar sands, oil shale, and coal. However, these cleaner resources will not appear overnight—nor will the vehicles that use them.

6.2.2. Key Challenges for Low-Carbon Fuel

Challenges

Making progress on producing new fuels for transportation has proved even more challenging than boosting vehicle efficiency—highlighted by the fact that oil and other petroleum products still account for 98 percent of fuel used for transportation. This hard road exists because new fuels face three barriers: technological, infrastructure-related, and behavioral.

6.2.2.1. Technology and Cost Hurdles

Whether the option is low-carbon biofuels, renewable electricity, or clean hydrogen, a transition to new fuels has stalled because either the fuel or the vehicle faces technological barriers to becoming widely available at a reasonable cost.

While creating vehicles that can run on biofuels is not a challenge, making the fuel at a competitive cost is. Producing cellulosic biofuels requires special en-

Tomorrow's plug-ins will be competing against conventional hybrids with fuel economy as high as 55 miles per gallon.

zymes that break down the walls of the plant cells. These enzymes are expensive today—both because they are still in development and because they are made on such a small scale. More research and development is needed to lower the cost, and demonstration programs are essential to start scaling up production to bring down costs. Similarly, the technology exists for biomass-to-liquid fuel, but demonstration projects are needed to scale up production to help bring down costs.

Unlike biofuels, making renewable electricity itself is less of a challenge. As Chapter 3 shows, electricity from wind can be cost-competitive with electricity from natural gas. Instead, the technological challenge of using electricity as a transportation fuel comes from the vehicle. Researchers have been trying for decades to develop batteries that are both durable and cost-effective, and several companies are working hard to reach this milestone. However, as evidenced by the projected

$40,000 price tag for GM's plug-in hybrid, more R&D is needed to cross the finish line.

As noted, the costs of plug-ins are expected to come down, making them less expensive to own than today's conventional vehicles. However, tomorrow's plug-ins won't be competing against today's cars. Instead, they will be competing against conventional hybrids with fuel economy as high as 55 miles per gallon.

A plug-in will cost an additional $2,800 over its lifetime compared with such a vehicle, assuming gasoline at $2.50 per gallon for half of all mileage, and electricity at $0.10 per kilowatt-hour for the other half. For the average plug-in owner to break even with the average hybrid owner, gasoline would have to reach $4.50 per gallon (or $3.50 per gallon for an owner with short commutes who can take advantage of all-electric operation three-quarters of the time). The high up-front

Shell Hydrogen and General Motors teamed up to build the first combined hydrogen and gasoline fueling station in North America, located in Washington, DC.

cost of plug-ins—combined with the fact that they pay for themselves only at higher gas prices or for owners with short commutes—poses a significant challenge.

Hydrogen, on the other hand, faces challenges in making both the fuel and the vehicles. Fuel cell vehicles require more R&D to lower costs, because of their extensive use of platinum and the cost of onboard storage of hydrogen. And although hydrogen can be produced cost-effectively from natural gas, further research is needed to cost-effectively produce hydrogen from renewable electricity or directly from sunlight. Hydrogen can also be produced from coal, but making that

an effective low-carbon option requires unproven technology, such as carbon capture and storage.

6.2.2.2. Infrastructure Barriers

A network of charging and alternative fueling stations is essential to the market success of low-carbon fuels and the vehicles that use them. As the number of vehicles that can run on low-carbon fuels grows, the infrastructure must grow as well.

Many biofuels, especially ethanol, require their own corrosion-resistant pumps and storage tanks at fueling stations, though regulations have already required such changes. Many of the nation's more than 160,000 gas stations already carry ethanol that has been blended into gasoline at low levels. However, fewer than 2,000 stations carry E85, a mixture of 85 percent ethanol blended with gasoline, which will be needed if ethanol is to grow to more than 10 percent of U.S. fuel use (EERE 2009c). Further, ethanol cannot be shipped in existing pipelines because of the risk of water contamination. Other biofuels will have to replace ethanol, or dedicated pipelines may be needed to supply larger amounts of biofuels.

Because plug-ins can operate on gasoline alone and have smaller battery packs than battery-electric vehicles, they will not require an extensive high-power charging infrastructure. Most overnight charging can occur at homes or businesses. However, such charging will require added equipment. A report from the U.S. Department of Energy (DOE) puts the cost of adding charging capabilities to a home at less than $900 (Morrow, Karner, and Francfort 2008).

Initially, battery-electric vehicles will be used by fleet operators or in urban locations, and will not require public charging stations. The cost of a charging system for such vehicles is higher than for plug-ins because quickly recharging the large battery pack requires more power. The DOE estimates that the cost of a residential charger is about $2,200 (Morrow, Karner, and Francfort 2008). If battery technology advances to a point where long-range electric vehicles do reach the market, the nation will need a public charging network.

As the numbers of plug-ins and battery-electric vehicles rise and they are integrated into the electricity grid, added costs and benefits will emerge. If large numbers of these vehicles increase electricity demand beyond today's levels, they will require investments by utilities and power producers.

The electricity grid will also need to tap into more renewable sources of power, or the environmental benefits of electric-drive vehicles will be limited. Standardized charging equipment and protocols, limits on charging during peak hours, and charging costs that vary with the time of day will also be critical to realizing benefits from widespread adoption of plug-ins or battery-electric vehicles.

Fuel cell vehicles will require the largest investment in infrastructure, because of the need to expand both the production and distribution of hydrogen. The National Research Council estimates that an aggressive goal of putting more than 60 million fuel cell vehicles on the road will require an investment of $8 billion from 2008 to 2023, and as much as $140 billion through 2035 (NRC 2008).

Regardless of the technology, widespread growth of clean vehicles will require both significant and intelligent investments in infrastructure.

6.2.2.3. Corporate and Consumer Behavior

Ensuring that the technologies for clean transportation fuels and vehicles work and that their costs are reasonable will probably still not be enough. The conventional gasoline car has been around for more than 100 years and is embedded in our way of life. Shifting to alternatives will require changes in the way we refuel our vehicles and changes in the vehicles themselves. For companies to invest in those changes, they need to believe that people will embrace them. But people will not do so unless they believe that the alternatives are viable.

The problem gets even worse when we consider petroleum prices. When they are low, the existing fuel supply industry resists competition, and consumers and the auto industry are less interested in alternatives. When petroleum prices are high, steering the fuel supply industry away from enormous short-term profits toward long-term alternatives will remain a challenge— just when consumers and the auto industry are most interested in those alternatives. And whether petroleum prices are high or low, almost everyone seems to resist fuel-pricing policies that support a shift to alternatives, despite their potential. All these challenges create a chicken-and-egg problem that can doom even cost-effective fuels.

6.2.3. Key Policies for Moving to Low-Carbon Alternatives

6.2.3.1. Making Them Available at the Pump

One approach is to put performance-based standards in place to make sure that low-carbon alternatives

Today, only about 4,000 out of the more than 160,000 U.S. gas stations carry biofuels, electricity, hydrogen, or natural gas. A strong policy push is needed to make these low-carbon fuels more widely available. A low-carbon fuel standard supports access to, and innovation in, transportation fuels while ensuring that these fuels contribute to both energy and climate security.

become available. The 2007 energy bill passed by Congress included the first federal policy requiring fewer heat-trapping emissions over the full life cycle of a fuel—including harvesting, production, delivery, and use, as well as direct and indirect emissions from the clearing of land for crops. This provision—the renewable fuel standard (RFS)—requires fuel providers to buy 36 billion gallons of biofuels by 2022, and sets low-carbon performance standards for at least 21 billion of those gallons. However, because the RFS does not regulate emissions from gasoline, biofuel production already in place, or other fuels, it covers only about 10 percent of the transportation fuel market. A more comprehensive approach—a low-carbon fuel standard that covers all transportation fuels—can protect and build on the benefits of the RFS.

A low-carbon fuel standard supports innovation in transportation fuels while ensuring that they contribute to both energy and climate security. Under such a standard, suppliers must reduce emissions from the fuels they sell on an average per-unit-of-energy (or energy intensity) basis. They can meet this requirement

TABLE 6.4. Potential of Advanced Vehicles and Fuels

	2020	2030
Low-Carbon Fuel Standard: Reduction in Carbon Intensity for All Transportation Fuels vs. 2005[a]	3.5%	10%
Sales of Advanced Light-Duty Vehicles Spurred by Regulations[b]	2.0%	20%

a This standard would require a reduction in life-cycle grams of CO_2 equivalent per Btu of all fuel used for transportation, including cars and light trucks, medium- and heavy-duty vehicles, rail, air, shipping, and other miscellaneous uses. If the standard is restricted to highway vehicles (cars, light trucks, and medium- and heavy-duty vehicles), the figure for 2020 would be 4.5 percent, and that for 2030 would be 14 percent.

b This represents the fraction of light-duty vehicles that are plug-in hybrids, or pure battery and fuel cell vehicles delivering equivalent benefits.

by blending low-carbon biofuels, such as cellulosic ethanol, into a fuel, or by improving refinery efficiency. Suppliers can also sell clean hydrogen or renewable electricity, or purchase credits from those that do. A low-carbon fuel standard requires providers to account for emissions from high-carbon alternatives—such as liquid coal, which has double the carbon emissions of gasoline—by either avoiding them or offsetting them with cleaner fuels.

Based on the potential to produce about 30 billion gallons of low-carbon biofuels, the nation should be able to reduce carbon emissions from all transportation fuels 8 percent by 2030. A reasonable goal of a 10 percent improvement in refinery efficiency can add another 1 percent cut in emissions. And if plug-in hybrids or other electricity-based vehicles account for 20 percent of sales of new vehicles, and use electricity producing 70–80 percent fewer carbon emissions than today (see below), such vehicles should cut the carbon intensity of fuels by another 1–1.5 percent.

All told, those technologies could cut carbon emissions from transportation fuels by about 3.5 percent in 2020, and 10 percent in 2030 (see Table 6.4).[67] California—which is leading the nation with a low-carbon fuel standard that aims to reduce emissions from transportation fuels 10 percent by 2020—would play a key role in fulfilling such a national standard (CARB 2009). California consumes 11–12 percent of the nation's transportation fuel, so its requirement would cut the nation's carbon emissions from such fuel by about

1.2 percent in 2020, partly through importing biomass and low-carbon fuels from other parts of the country (EIA 2009e).

A nationwide low-carbon fuel standard is constrained by how quickly technologies such as cellulosic biofuels and biomass-to-liquids can scale up, and by the amount of land available to produce the needed biomass. Conservative assumptions in our Blueprint case about the availability of land for producing biomass for transportation fuel mean that such fuel would not have significant direct and indirect carbon emissions. More optimistic assumptions—such as those in a 2007 report from the DOE's Energy Information Agency—suggest that annual biofuel production could reach 45–60 billion gallons, which could support a 15 percent low-carbon fuel standard in 2030 (EIA 2007).

6.2.3.2. Moving Consumers to the Cleanest Fuels
All these challenges point to the need for significant research, development, and deployment programs to speed low-carbon fuels to market. While industry will bear much of the cost of such programs, and pass them on in the prices of vehicles and fuels, these technologies are both risky and important enough to suggest a clear role for government-funded programs. Those will need to support everything from basic research to grants for the pilot plants needed to prove the technology and begin the process of scaling up production.

But even R&D will not be enough to overcome the chicken-and-egg problem, especially when it comes to the more advanced vehicles like plug-ins, which will cost even more than conventional hybrids, and fuel cell vehicles. Even a low-carbon fuel standard is unlikely to spur early widespread use of the best technologies, if refiners find simpler alternatives. While that may be fine in the near term, it will further delay progress on these technologies, and may compromise their ability to bear fruit in the longer term.

Subsidies in the early years, when costs are high, can lead advanced fuels and vehicles to the market. One study put the incremental cost of bringing hydrogen fuel to market—and the number of fuel cell vehicles to about 5 million by 2023—at more than $50 billion (NRC 2008). Plug-ins may require a similar level of funding, to cover the extra up-front costs of more advanced vehicles. These resources could come from

---

67 If the low-carbon fuel standard applied only to highway vehicles—cars and light trucks, and medium- and heavy-duty vehicles—the equivalent values would be a 4.5 percent reduction in carbon emissions from highway fuels in 2020, and a 13.7 percent cut by 2030.

auctions of carbon allowances under the cap-and-trade program, or as part of a broader effort to create green jobs in the coming decade.

Another alternative is to create a performance-based version of California's Zero Emission Vehicle program. While that program has encountered delays, direct requirements for advanced vehicles have spurred the development of hybrids and significant progress on battery and fuel cell technology (Turrentine and Kurani 2000).

6.3. The Road Less Traveled: Reducing Vehicle Miles

The nation can make great strides in improving vehicle efficiency and producing cleaner fuels, but technology alone will not keep pace with growing demand for personal and freight travel if we continue on our current path. The classic suburban American lifestyle is predicated on driving a personal car a growing number of miles.

Since 1980, the number of vehicle miles traveled (VMT) in cars and light trucks has grown three times faster than the U.S. population, and nearly two times faster than vehicle registrations (Ewing et al. 2007). VMT is expected to grow at a slower pace than historical trends from 2005 to 2030 but continue its upward trajectory, growing nearly two times faster than the U.S. population (EIA 2008a).[68]

While today's fleet of cars and light trucks travels about 2.7 trillion miles a year, that number could easily reach 3.8 trillion miles by 2030—a 42 percent increase (1.4 percent per year) (CTA 2008; EIA 2008a). Freight travel could rise by a similar amount, while air travel could grow by as much as 60 percent (2.2 percent per year) (EIA 2008a).

To slow growth in vehicle miles traveled, the nation needs to promote compact development, provide drivers with market-based incentives to drive less (such as pay-as-you-drive insurance and congestion mitigation fees), and give freight operators tools to increase the number of tons they haul per mile. Better-planned and more compact development can shorten car trips, increase the use of public transit and light rail, and provide substantial health and other benefits. By co-locating housing with jobs, improving access to walking, biking, and public transit, and revitalizing city centers, the road less traveled can promote healthy, vibrant, and desirable communities while cutting carbon emissions.

6.3.1. Potential for Reducing Car and Truck Travel

Potential

Overturning today's car-centric culture will require us to overhaul how and where we live and work. The technical potential to reduce travel is vast, although more research is needed on the cost and effectiveness of various strategies.

The next sections explore the technical potential for building smarter cities, reducing VMT through personal choice, raising the number of people using each vehicle and public transit, and moving goods more efficiently. Our analysis does not include the technical potential to reduce air, marine, and off-road VMT, or to improve transit options, as in providing high-speed electric rail.

6.3.1.1. Smarter and More Compact Development

More compact and better-planned cities could reduce VMT by up to 30 percent (Ewing et al. 2007). One study found that residents of sprawling, suburban Atlanta and Raleigh drove more than 30 miles per person per day, while residents of compact cities such as Boston and Portland drove fewer than 24 miles per person

Growing in popularity, car sharing greatly reduces vehicle miles traveled simply by reducing the number of vehicles on the road. The car-sharing company Zipcar offers trucks, compacts, and hybrids for urban transport; members can reserve a car months or minutes in advance. Instead of purchasing a pickup truck and driving it year-round, members can rent trucks just for the times they need to transport heavy loads.

---

68   This analysis of VMT is based on the EIA's high-gas-price scenario ($3.50 per gallon) versus its baseline scenario ($2.50 per gallon) in its *2008 Annual Energy Outlook*. Higher gas prices suppress VMT for light-duty vehicles from the baseline of 1.7 percent per year to 1.4 percent per year.

Over the next 20 years there are significant opportunities to shape new growth in ways that enable denser and more livable communities. As downtowns are revitalized, areas can be reserved for walking, biking, and public transit. New communities can have housing, shopping, parks, and jobs integrated with public transit.

per day—more than 25 percent less (Ewing, Pendall, and Chen 2002). Another study found that residents of compact regions drive up to one-third fewer miles than the U.S. average (Bartholomew 2005 and 2007 in Ewing et al. 2007).

Construction of new buildings and revitalization of existing neighborhoods provide the greatest opportunity to capitalize on smart growth to reduce VMT. According to one study, "By 2030, about half of the buildings in which Americans live, work, and shop will have been built after 2000" (AASHTO 2007). That means that the next 20 years will provide significant opportunities to shape new growth in ways that enable denser and more livable communities. As we revitalize our downtowns, we can reserve space for walking, biking, and public transit, while new communities can integrate housing, shopping, parks, and jobs with transit. These approaches would feed a smart-growth revolution that entails rethinking how we move people and goods.

6.3.1.2. Choosing to Drive Less

For discretionary trips, drivers can make a conscious choice to reduce VMT, spurred by good intentions, personal preferences (such as the desire to avoid time in traffic or trapped behind the wheel of a car), or market-based incentives. Given historical trends, good intentions and personal preferences alone are unlikely to provide much of a reduction. However, market-based incentives have proven successful.

The higher the cost per mile of driving, the more likely a consumer will be to drive less. A recent study of 84 U.S. urban areas from 1985 to 2005 found that for every 1 percent increase in the price of fuel, VMT fell by 0.17 percent (Ewing et al. 2007). A study by Cambridge Systematics found that pay-as-you-drive insurance—which charges drivers based on how many miles they drive—could reduce national VMT more than 7 percent (Cowart 2008).[69]

6.3.1.3. Bring a Friend: Raising the Occupancy of Personal Vehicles

One simple way to reduce travel is to increase the number of people riding in each vehicle, reversing historical trends. The average U.S. vehicle carries 1.6 people, and occupancy drops to barely more than one person (1.1) during trips to work (CTA 2008).

Carpooling is the key to increasing the occupancy of personal vehicles. If people who drive to work carpooled with just one other person (HOV-2) every other day, annual car and truck travel would drop by more than 5 percent, and average vehicle occupancy during work trips would rise to about 1.4.[70] If commuters carpooled with two other people (HOV-3) for about 60 percent of work trips, annual travel would fall about 10 percent, and average occupancy during work trips would rise to about 1.8.

6.3.1.4. Ride the Bus: Expanding Ridership on Public Transit and in Vanpools

Urban and suburban areas need greater access to public transportation and vanpools to help cut carbon emissions. As of 2001, less than one-third of the U.S. population lived within about a block of a bus line, and only about 40 percent lived within a half-mile (NCTR 2007). The situation is even worse for rail: only about 10 percent of the U.S. population lived within a mile of a rail stop, while only about one-quarter lived within five miles (NCTR 2007). As a result of low ridership, buses release more global warming emissions per passenger-mile than cars.

---

69   This figure does not account for induced demand from reduced congestion, which will offset some of the gains of pay-as-you-drive insurance.

70   Travel to work would shrink by about 25 percent. However, such travel represents only 27 percent of all VMT for cars and light trucks, so the overall impact of such a shift is much smaller (CTA 2008).

Public transportation advocates have pointed to the potential to at least double the capacity and ridership of bus and rail transit by 2030, at an annualized cost of about $21 billion (AASHTO 2007). That would

By carpooling with a co-worker every other day or shifting to a four-day work-week, Americans can reduce their personal travel by more than 5 percent.

represent an important start in satisfying Americans' awakening appetite for public transit (Sun 2008). Individual drivers who switch from a car or SUV to vanpools, bus, or electric rail cut their carbon footprint significantly, with the benefits rising as more people switch.

The U.S. mass transit system is so small that doubling it will reduce VMT by only about 2 percent, given today's ridership levels. Expanding ridership could boost this impact significantly, especially in regions that use transit least effectively. The nation clearly needs to make major investments in public transit, but such investments will bear most of their fruit after 2030.

6.3.1.5. Working Up a Sweat or Working from Home: Near-Zero-Carbon Options

We can cut carbon emissions dramatically by replacing car trips with walking or biking. However, in 2000 fewer than 3 percent of Americans reported walking to work, while less than one-half of 1 percent reported bicycling to work (CTA 2008).

Flexible workplace policies that allow employees to work at home or shift to four days per week at 10 hours per day can also reduce car and truck use. In 2000, slightly more than 3 percent of Americans reported working from home (CTA 2008). By shifting to a four-day workweek or working from home one day per week, the typical American could cut his or her overall travel by about 5 percent. By working at home two days a week, he or she could cut annual travel by about 10 percent.

New technology and expanded transit service can increase ridership while dramatically cutting emissions and creating jobs. A number of Indiana cities, for example, have added hybrid buses to their transit fleets, including Indianapolis (pictured here), Muncie, Fort Wayne, Bloomington, greater Lafayette, Evansville, and Terre Haute. At least three Indiana-based companies manufacture hybrid bus drivetrains or components.

Changing our daily commutes by walking, biking, or telecommuting can dramatically cut carbon emissions. Unfortunately, less than 10 percent of Americans report taking advantage of these alternatives. This statistic highlights the need for policies and funding to make these commuting options safe and accessible— true alternatives to being stuck in traffic alone in a car.

Working from home is not quite carbon-free, because the use of lights, computers, heating, and cooling does grow. However, avoiding the use of those resources at the office should offset much of that use. Working longer hours but fewer days is not quite carbon-free for the same reason, and also because that practice could encourage people to take more leisure

or shopping trips on their extra day off. However, those practices are a good start.

6.3.1.6. Car and Truck Travel: All of the Above

As with cleaner vehicles and fuels, no silver bullet can preserve the mobility we enjoy today while reducing overall travel. Instead, the nation will have to pursue a variety of approaches to ensure that people live productively while relying less on personal vehicles.

A recent analysis by Cambridge Systematics points to the potential for significant reductions in projected car and light-truck travel (Cowart 2008). That study evaluated a suite of policies to reduce VMT through more compact communities, per-mile pricing policies, and other smart-growth strategies. The study found that these approaches could reduce the annual growth rate in VMT for light-duty vehicles from 1.7 percent to 0.9 percent.[71] Part of that reduction could come from a doubling of transit.

6.3.1.7. Shifting Freight Back to Rail

Just as consumers can shift to more efficient transportation modes such as transit, biking, and walking, companies can also shift to rail as a more efficient mode for moving goods than trucks.

Moving goods by rail is about five times more efficient than doing so by truck, based on weight, primarily because rail transports dense, heavy cargo such as coal.[72] However, even for lighter-weight loads more typical of 18-wheelers, trucks emit two to three times more carbon emissions than trains (Mathews 2008). And rail is likely to retain that advantage over trucks during the coming decades, although it may erode if improvements in truck efficiency outpace those in rail.

Estimates from the American Association of State Highway and Transportation Officials (AASHTO) indicate that 1 percent of truck freight could shift to rail by 2020, and an analysis by two national laboratories points to the potential for a 2–5 percent shift (AASHTO 2003; IWG 2000). A conservative estimate is that about 1.5 percent of freight could move from trucks to rail by 2020, and at least 2.5 percent could shift by 2030.

---

71 The analysis showed an 18 percent reduction in projected light-duty travel in 2030 of more than 4 trillion miles, accounting for induced demand. The baseline projection of a growth rate of 1.7 percent per year is higher than that used in our study (1.4 percent per year). Our figure means that a reduction to 0.9 percent per year should be even easier to achieve.

72 The average value of five times is based on data from the *Transportation Energy Data Book* (CTA 2008), and assumes that freight trucks carry 11.8 tons per mile, based on statistics from the U.S. Department of Transportation (U.S. DOT 2000). Using the most common type of tractor-trailer (van trailers), we found the average payload to be 30,555 pounds. The loads of such combination vehicles are the most likely to shift to rail. Using the U.S. Census Bureau's 2002 Vehicle Inventory and Use Survey Microdata for Class 8 trucks, we determined that their "empty miles" averaged 23 percent.

6.3.2. Key Challenges for Smarter Travel, Freight Transport, and Cities

6.3.2.1. Lack of Funding

There is no denying that expanding transit costs money, and that finding those funds is a challenge. Estimates show that annual funding for highways and transit falls about $10 billion short of what the nation needs just to maintain the existing system. Closing that gap would require a 6-cent-per-gallon increase in diesel and gasoline taxes, while doubling transit capacity would require another 12 cents per gallon. Together those increases would double today's 18-cent-per-gallon gasoline tax—a drop in the barrel compared with price swings in 2008, but likely still a significant political hurdle. Making matters ironically worse, rising fuel economy will cut projected gasoline and diesel use, expanding the funding shortfall.

If businesses shifted 2.5 percent of goods now shipped by truck to rail by 2030, carbon emissions from those shipments would drop by half or more.

6.3.2.2. The Impact of a Lower Cost of Driving

Blueprint policies that require new vehicles to reduce carbon emissions will help push fuel economy to 50–55 miles per gallon in 2030 and reduce the per-mile cost of driving—potentially giving consumers an incentive to drive even more. As the cost of driving falls, consumers have less incentive to carpool, take public transit, or explore near-zero-carbon options such as biking and walking. Increasing the efficiency of our car and truck fleets is essential. However, if doing so encourages people to drive more while still saving money, it could dilute some of the carbon benefits of more efficient vehicles.

A similar impact may result from reducing the number of vehicles on the road, as that will lower another cost: time wasted in congested traffic. That may encourage people who avoid congested routes to switch to those routes, again reversing some of the progress.

6.3.2.3. Weak Market Signals

Funding challenges are directly tied to the fact that consumers do not actually pay the full costs of driving.

Given that Americans are not paying enough in gasoline taxes to maintain today's highway system, they are clearly not directly paying the full cost of the U.S. reliance on personal vehicles, including the national security costs of our dependence on oil, the health impacts from smog and toxic pollution, time lost owing to congestion, and the health and economic impacts of global warming, just to name a few. If we are not directly paying the full price of a resource, we are going to use too much of it, and will be less willing to switch to the many alternatives.

Transporting coal from where it is mined to where it is burned in a power plant takes up about 40 percent of our nation's rail capacity. Because rail is a more efficient and less carbon-intensive way to ship freight, one surprising benefit of phasing out coal-fired power plants is that rail capacity would be freed up to handle other kinds of freight currently transported in trucks, helping to reduce the global warming impact of the shipping sector.

BOX 6.4.

SUCCESS STORY

It Takes an Urban Village to Reduce Carbon Emissions

Arlington, VA, has won national awards for its "urban village" model of smart growth.

Chicago's Climate Action Plan includes a smarter, energy-efficient transportation mix.

The trolley carried its first passengers from Clarendon, VA, across the Potomac into Washington, DC, in 1896 (APA 2007). It served not only commuters but also shoppers, transporting them to stores along its lines. Today the "urban villages" along the old trolley lines—Clarendon, Courthouse, Ballston, and others—make Arlington County one of the most desirable communities in the metropolitan DC area.

Although there are no more streetcars, the spirit of the trolleys is alive in Arlington County. In contrast to its suburban cousins in Maryland and northern Virginia, the county used its rail and bus system as a foundation for smart growth, encouraging business development while preserving unique neighborhoods.

Under its General Land Use Plan, Arlington concentrated dense, mixed-use development around its Metro stations beginning in the mid-1980s. These urban villages emphasize pedestrian access, promote safety through traffic "calming," provide bike lanes, and create highly desirable living spaces by incorporating public art, pocket parks and street trees, wide sidewalks with restaurant seating, and street-level retail (EPA 2002).

While much of the nation followed the trajectory of urban sprawl, Arlington County boasts 22,500 apartments and condos, townhouses, and single-family detached homes, as well as a thriving commercial base (EPA 2002). Mindful of the area's socioeconomic disparities, county government and civic groups worked to spread the benefits equitably among all residents. Affordable Housing Protection Districts, for example, help preserve low- and moderate-income apartment units (CPHD 2008a).

Metro ridership in the corridor doubled between 1991 and 2002. And to expand residents' access to public transportation, the county created the Arlington Rapid Transit

Arlington County's urban villages emphasize pedestrian access, promote safety through traffic "calming," provide bike lanes, and create highly desirable living spaces by incorporating public art, pocket parks and street trees, wide sidewalks with restaurant seating, and street-level retail.

system (ART)—a fleet of 30 smaller, handicapped-accessible buses that can navigate neighborhoods and are well integrated into the comprehensive network of bus and train lines in the nation's capital and the surrounding region (ART 2009).

The resulting health, environmental, and other quality-of-life benefits are equally impressive. Almost half of Arlington residents use transit to commute (APA 2007), while another 6 percent walk to work, compared with 2.5 percent nationwide (CPHD 2008b; Reuters 2007). Nearly 20 percent of county residents do not even own a car.

Heavily traveled Wilson Boulevard saw traffic drop nearly 16 percent from 1996 to 2006 (APA 2007). Commute time in Arlington County is the region's lowest, and both carbon and smog-forming emissions have fallen dramatically. The county has accomplished all this while maintaining a high level of municipal services and the lowest property tax rate of any jurisdiction near the nation's capital (CPHD 2008b).

The urban village model has won national awards for smart growth (EPA 2002), and the American Planning Association recently showcased Arlington's main corridor as one of the Great Streets of America (APA 2007).

Meanwhile other cities are forging their own smart-growth paths. Chicago's Climate Action Plan places strong emphasis on transit-oriented community growth. Atlanta has focused on urban renewal through its downtown Atlantic Station project (EPA 2005). And the outer-rim suburb of Buckeye, AZ, near Phoenix, is working to become its own bedroom and business community (Suarez 2008). Whatever the approach, a commitment to sustainable growth is one way to help us reach our lower-carbon future.

6.2.2.4. Incentives for Unrestricted Growth

A number of existing laws actually encourage sprawling development, which requires greater use of cars. For example, zoning requirements that do not allow commercial and residential uses to mix limit the potential to integrate transit, housing, and shopping. Local ordinances that require taxpayers to fund the expansion of utilities to new houses and businesses ensure that developers do not pass on the full costs of building outside existing communities. And formulas for distributing federal highway funds that focus on expanding roadways rather than mitigating traffic further encourage sprawl.

6.3.3. Key Policies That Provide New Options for Getting There from Here

Policies

As U.S. history shows, the barriers to reducing projected vehicle miles traveled are anything but trivial. However, several public policies could help overcome these barriers.

6.3.3.1. Smart-Growth Policies

The biggest job of all is rethinking and reinventing where we live. Much of this work has to happen at the local level, such as by changing zoning laws to allow more mixed use, and requiring developers to pay the full costs of extending utilities to their projects. However, the federal government can help move these approaches along through a variety of steps.

Agencies should tie existing and future highway funding to performance metrics—whether cuts in carbon emissions or vehicle miles, or more efficient use of infrastructure. Highway funds represent a significant transfer of taxpayer dollars, and their use should focus on delivering public benefits. The nation should also reform the home mortgage tax deduction, to allow higher deductions for homes near transit or in mixed-use developments.

6.3.3.2. Pay as You Drive

Another straightforward approach to overcoming barriers is to require that people pay the actual costs of their daily driving. The initial response to asking people to pay more for every mile they drive is resistance, as they see that approach as raising their expenses. However, that is just a misunderstanding.

Americans are already bearing those costs. For example, we pay higher hospital bills and health insurance rates because of asthma and lung disease stemming from smog. We also pay higher income taxes to help secure U.S. access to oil, and to cover the shortfall in road repair funds. And if we do not cover these costs elsewhere, we still pay them through spikes in gas prices, more costly car repairs, and time wasted owing to congestion, potholes, bridge collapses, and road closures. Consumers and businesses that pay driving fees will therefore see tax cuts elsewhere, lower health care costs, fewer price spikes, and less congestion.

We can start with two key measures: highway user fees and pay-as-you-drive insurance.

Highway user fees. As the fuel economy of vehicles rises, people will use less fuel, save money, and pollute less. But because they use less fuel, they will also pay fewer gasoline taxes, which are collected on a

TABLE 6.5. Potential for Reducing Vehicle Miles Traveled

Potential for Reducing Vehicle Miles Traveled	2020	2030
Assumed Policy Impact: Reduction in Annual Growth in Vehicle Miles Traveled (VMT)[a]		
Light-Duty Vehicles[b]	Reduce growth in VMT from baseline of 1.4% per year to 0.9% per year	
Trucks[c]	Reduce VMT by 0.1% per year, on top of all other policy effects	
Policies and Costs for Light-Duty Vehicles		
Transit[d]	Ramp up transit funding to reach $21 billion per year by 2030	
Pay as You Drive		
Highway User Fee 1: Maintain Existing Funding Levels[e]	$0.005 per mile	$0.011 per mile
Highway User Fee 2: Congestion Mitigation Fee Used to Fund Transit[d]	$0.004	$0.006
Total User Fees	$0.009 per mile	$0.017 per mile
Pay-as-You-Drive (PAYD) Insurance[e]	$0.07 per mile	$0.07 per mile
Federal Funding for PAYD Pilot Programs	$3 million per year for 5 years	
Tax Credit for PAYD Electronics	$100 million per year for 5 years	
Smart Growth[f]	$0.00	$0.00
Policies and Costs for Heavy-Duty Vehicles		
Switch from Truck to Rail[g]	$0.00	$0.00

Notes:

a NEMS is unable to model the full suite of policies needed to address vehicle travel. Instead, we inserted the total reductions in vehicle miles traveled that could result from such policies into UCS-NEMS.

b For the potential to reduce VMT from light-duty vehicles, we relied primarily on a recent analysis by Cambridge Systematics (Cowart 2008), which found that growth in light-duty VMT could be reduced to 0.9 percent per year.

c To evaluate the potential to reduce VMT from freight trucks, we assumed that policies can shift 2.5 percent of truck VMT to rail, based on potential highlighted in AASHTO 2007 and IWG 2000. This represents about a 0.1 percent annual reduction in freight truck travel. Actual freight truck travel will fall further as the economy shifts due to other policies, such as a cap-and-trade program and reduced oil use from higher vehicle efficiency.

d The congestion mitigation fee provides this funding, so we did not count it as a cost above that fee.

e Blueprint policies do not include these fees as a cost, because the Reference case would also need to raise the highway funding to pay for repair of existing roads, and would include the cost of insurance. Actual insurance costs would probably drop, because people would drive less under the Blueprint.

f Smart-growth policies could actually reduce costs, so we assumed that they are cost-neutral.

g Switching from truck to rail will likely entail some costs, but evaluating them was beyond the scope of our study.

Riding a bike to work is a healthy and affordable alternative to driving, and it saves time, too—no more wasting time stuck in traffic or circling parking lots.

per-gallon basis. Rather than raising those taxes to compensate, we should adopt a per-mile user fee that at least covers any resulting federal, state, and local shortfalls. We should also institute a congestion and air-quality mitigation fee that will at least cover the costs of doubling transit by 2030. The former would require at least a one-cent-per-mile road user fee by 2030. The latter would require a little more than a one-half-cent-per-mile fee—for a total of $0.009 per mile by 2020, and $0.017 per mile by 2030 (see Table 6.5).

Per-mile highway user fees do not represent a new cost to drivers, as the nation would need to raise the funds to maintain our roads in any case. And unlike income or sales taxes, such fees will have the added benefit of reducing the number of miles we drive. We therefore should not count these specific per-mile highway user fees as a "cost" of cutting carbon emissions.

Pay-as-you-drive insurance. Another cost that today's drivers already bear is car insurance. The price of that insurance is usually not tied to the number of miles we drive, despite the fact that the more we drive, the more we risk the accidents that insurance covers.

If insurance were tied to the number of miles we drive, the roughly $800 per year in insurance we pay would equal about $0.07 per mile—the equivalent of raising gas taxes by about $1.40 per gallon.[73] Two recent reports point to the potential of this approach to cut VMT by 7–9 percent (Bordoff and Noel 2008; Cowart 2008). The Blueprint analysis estimates an impact on the order of 5–6 percent (likely because of a higher per-mile baseline, reflecting higher gasoline costs).

Pay-as-you-drive (PAYD) insurance is more equitable, given that low-mileage drivers now subsidize high-mileage drivers, and consumers will save money

---

73   The cost of insurance is from Bordoff and Noel 2008. The per-mile figure is based on 11,500 miles per year. The per-gallon figure is based on 20 miles per gallon.

BOX 6.5.

Technologies and Other Options on the Horizon: Transportation

Breakthroughs in algae biofuels

High-speed rail

- **Dramatic expansion of all-electric cars and trucks.** If cost and other key hurdles are overcome, by 2050 most cars and light trucks could run on batteries or fuel cells supplied by renewable energy, effectively eliminating those vehicles as a significant source of carbon emissions. Many medium-duty and heavy-duty vehicles could also follow this path.

- **Advanced high-strength materials.** Carbon fiber, now used in aircraft, could become cost-effective for use in highway vehicles over the long term. Such uses would dramatically cut the weight—and thus the carbon emissions—of cars and trucks of all shapes and sizes while also increasing their safety.

- **Breakthroughs in third-generation biofuels.** From algae to efficient microbes that can digest almost anything, hoped-for breakthroughs could produce large volumes of liquid fuels with minimal land use.

- **High-speed and zero-emission rail.** Trains that can move rapidly between major cities while running on renewable electricity could replace airplanes for shorter trips, eliminating carbon emissions. Such a train system could also help shift freight from truck to rail, significantly reducing emissions from freight shipments.

- **Expanded transit-oriented development.** Cities do not expand overnight. With advanced planning, more and more cities could expand around transit, integrating homes, shopping, and transportation with parks and open areas across the country.

To double capacity and ridership for bus and rail transportation by 2030, more money must be set aside to expand and improve public transit. Chicago leads the way with an aggressive climate action plan (online at *www.chicagoclimateaction.org*) to reduce its emissions across various sectors including transportation (which currently comprises 21 percent of its total emissions). By investing in transportation alternatives such as public transit, bicycling, walking, car sharing, hybrid buses, and smart growth, Chicago can meet its goal of creating a convenient and energy-efficient transportation mix.

as reduced travel means fewer accidents and thus lower costs. Bordoff and Noel estimate that this approach could save the nation $32 billion just by reducing the number of accidents—or about $150 per vehicle, of which $34 per vehicle could accrue to insurers.

The way PAYD is implemented is also important. While it could be based on annual odometer readings, a once-per-year payment or rebate might not have the same impact as more immediate feedback. A better alternative is to install a GPS-based device to track mileage, which Bordoff and Noel estimate would cost $100 per vehicle, and to require periodic payment of insurance premiums. GPS technology could even allow us to pay for insurance along with a fuel purchase, combining pay-at-the-pump with PAYD.

Bordoff and Noel point to the cost of this device as a significant hurdle, because insurance companies would not save enough to cover it, so they might be unwilling to advocate for it. To overcome that hurdle, those analysts recommend a $100 tax credit per vehicle for insurance companies, for the first 5 million vehicles. That approach would put systems in place that could also support per-mile road user fees. Bordoff and Noel also recommend that the federal government spend $3 million per year for five years to establish pilot programs, and that states adopt laws clearing the way for PAYD insurance.

6.3.3.3. More Funding for Transit

If the nation is to double transit by 2030, we must set aside more money to expand and improve bus and rail transportation. Based on AASHTO recommendations, such a doubling would require additional dedicated funding that would reach $21 billion per year in 2030.

CHAPTER 7

We Can Do It: Analyzing Solutions to Global Warming

This chapter presents the results from implementing the Blueprint—a comprehensive suite of climate, energy, and transportation policies that tackle most sources of heat-trapping emissions in the electricity, residential, commercial, industrial, and transportation sectors, and that also allow a limited amount of offsets based on storing carbon in the agriculture and forest sectors (see Box 7.1).

Using the UCS-NEMS model and other analyses, we compared the impact of the Blueprint to that of a Reference case that assumes no new federal and state policies beyond the existing ones.[74] We also analyzed a No Complementary Policies case, which investigated the impact of stripping out all the sector-based complementary policies, and compared that case with the Blueprint case. (See Chapter 2 for more information.)

Our results include carbon prices and revenues under a cap-and-trade program, changes in energy use by fuel and sector, improvements in energy security (through reduced oil imports and a more diverse energy mix), and costs and benefits to consumers and businesses (see page 127).

Overall, our analysis shows that the Blueprint achieves significant cuts in net U.S. heat-trapping emissions in a timely manner while saving consumers and businesses significant amounts of money.

FIGURE 7.1. Net Cuts in Global Warming Emissions under the Climate 2030 Blueprint

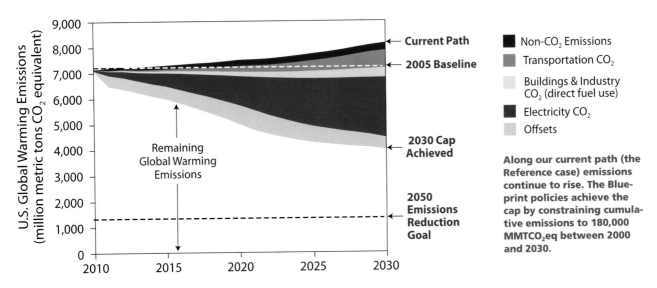

Along our current path (the Reference case) emissions continue to rise. The Blueprint policies achieve the cap by constraining cumulative emissions to 180,000 MMTCO$_2$eq between 2000 and 2030.

74   The Reference case includes policies that had become law by October 2008. The Reference case does not include the impact of the American Recovery and Reinvestment Act because it was passed after that date. However, the Reference case does include the (significant) impact of the 2007 Energy Independence and Security Act, as well as the effects of a variety of state renewable energy standards and the existing nuclear loan guarantee program.

Climate 2030 Blueprint Policies[a]

Climate Policies

Economywide cap-and-trade program with:

- Auctioning of all carbon allowances
- Recycling of auction revenues to consumers and businesses[b]
- Limits on carbon "offsets" to encourage "decarbonization" of the capped sectors
- Flexibility for capped businesses to over-comply with the cap and bank excess carbon allowances for future use

Industry and Buildings Policies

- An energy efficiency resource standard requiring retail electricity and natural gas providers to meet efficiency targets
- Minimum federal energy efficiency standards for specific appliances and equipment
- Advanced energy codes and technologies for buildings
- Programs that encourage more efficient industrial processes
- Wider reliance on efficient systems that provide both heat and power
- R&D on energy efficiency

Electricity Policies

- A renewable electricity standard for retail electricity providers
- R&D on renewable energy
- Use of advanced coal technology, with a carbon-capture-and-storage demonstration program

Transportation Policies

- Standards that limit carbon emissions from vehicles
- Standards that require the use of low-carbon fuels
- Requirements for deployment of advanced vehicle technology
- Smart-growth policies that encourage mixed-use development, with more public transit
- Smart-growth policies that tie federal highway funding to more efficient transportation systems
- Pay-as-you-drive insurance and other per-mile user fees

a See Chapters 3, 4, 5, and 6 for more details on these policies.

b We could not model a targeted way of recycling these revenues. The preferred approach would be to target revenues from auctions of carbon allowances toward investments in energy efficiency, renewable energy, and protection for tropical forests, as well as transition assistance to consumers, workers, and businesses moving to a clean energy economy. However, limitations in the NEMS model prevented us from directing auction revenues to specific uses. Instead, we could only recycle revenues in a general way to consumers and businesses.

7.1. The Reference Case: Significant Growth in Carbon Emissions

In the Reference case, U.S. global warming emissions rise from 7,181 million metric tons carbon dioxide equivalent (MMTCO$_2$eq) in 2005 to 8,143 MMT-CO$_2$eq in 2030—an increase of 13.4 percent. Total U.S. energy use rises by nearly 16 percent over the same period, or an average of 0.74 percent per year, with fossil fuel use growing 10 percent.

Most of the increase in carbon emissions and energy use in this scenario stems from greater use of coal to generate electricity and produce liquid fuels for the transportation and industrial sectors. Growth in the use of natural gas in industry and buildings also makes a modest contribution to rising carbon emissions.

The use of oil and other petroleum products declines in the Reference case, as policies in the 2007 Energy Independence and Security Act improve the efficiency of vehicles and expand the use of biofuels. The nation's reliance on renewable energy from wind, solar, geothermal, and biomass resources more than triples by 2030 under the Reference case. Contributions from nuclear energy and hydropower remain relatively flat. However, overall, the nation continues to rely heavily on both fossil fuels and nuclear power to provide 89 percent of its energy.

7.2. The Big Picture: The Blueprint Cuts Carbon Emissions, Saves Money, and Reduces Energy Use

7.2.1. Significant Near-Term and Medium-Term Cuts in Emissions

Under the Blueprint, the nation achieves significant near-term and mid-term cuts in global warming emissions at a net savings to consumers. Blueprint policies reduce U.S. carbon emissions enough to meet a cap set at 26 percent below 2005 levels in 2020, and 56 percent below 2005 levels in 2030 (see Figure 7.1).

In Figure 7.1, the actual year-by-year trajectory of cuts in emissions differs from the trajectory specified under the cap-and-trade program, because that program gives companies the flexibility to bank extra carbon allowances in early years and withdraw them in later years.[75] However, cumulative heat-trapping emissions from 2000 to 2030 remain the same under both

---

75 See Section 7.3.2 for a fuller explanation of how the banking and withdrawing occurred in our results. Further information is available in Appendix B online.

Major Findings of the Climate 2030 Blueprint

IN 2030, THE UNITED STATES CAN:

#1 Meet a phased-in cap on global warming emissions representing a 56 percent drop from 2005 levels, at a net annual savings of $464 billion to consumers and businesses.

7,181 MMTCO$_2$eq

Pre-cap Emissions

Blueprint Cap

3,145 MMTCO$_2$eq

2005   2030

#2 Reduce annual energy use by one-third compared with the Reference case.

#3 Cut the use of oil and other petroleum products by 6 million barrels per day compared with 2005, reducing imports to less than 45 percent of our needs and cutting projected expenditures on those imports by more than $85 billion, or more than $160,000 per minute.

#4 Reduce annual electricity generation by 35 percent compared with the reference case, through the use of greater energy efficiency in buildings and industry, while producing 16 percent of the remaining electricity with combined heat and power and 40 percent with renewable energy resources, such as wind, solar, geothermal, and bioenergy.

#5 Rely on complementary policies to deliver cost-effective solutions based on efficiency, conservation, and renewable energy. Excluding Blueprint policies in the energy and transportation sectors reduces net cumulative consumer and business savings through 2030 from $1.7 trillion to $0.6 trillion.

FIGURE 7.2. The Source of Cuts in Global Warming Emissions in 2030
(Blueprint case vs. Reference case)

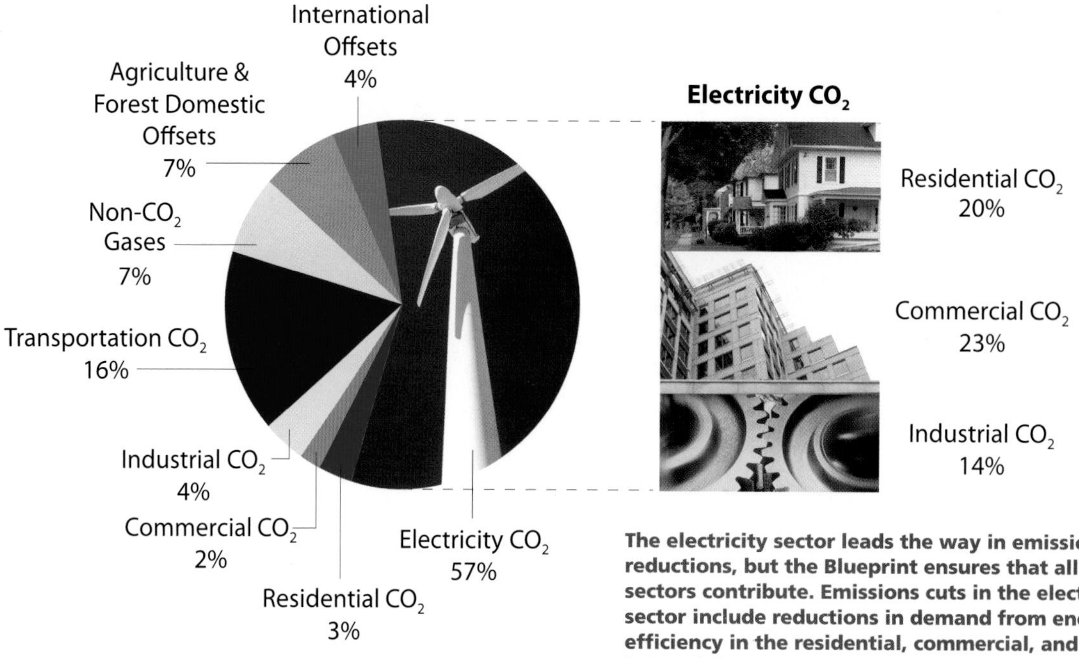

The electricity sector leads the way in emissions reductions, but the Blueprint ensures that all sectors contribute. Emissions cuts in the electricity sector include reductions in demand from energy efficiency in the residential, commercial, and industrial sectors.

Note: Refinery emissions have been allocated to the appropriate end-use sector. Transportation emissions do not include full well-to-wheel emissions, because UCS-NEMS does not account for emissions associated with products imported into the United States.

trajectories: about 180,000 MMTCO$_2$eq.[76] If the nation continues along the path of the cap modeled here, we could remain in the mid-range of the U.S. carbon budget in 2050 (165,000–260,000 MMTCO$_2$eq from 2000 to 2050), with cumulative emissions of 216,000 MMTCO$_2$eq by 2050.

Under the Blueprint, actual emissions are 30 percent below 2005 levels in 2020, and 44 percent below 2005 levels in 2030. Those reductions are 33 percent below those of the Reference case in 2020, and 51 percent below the Reference case in 2030 (see Figure 7.1). These reductions are a first and critical step to putting the nation on a path to achieving the 2050 targets needed to avoid the most dangerous effects of climate change.

In 2030, the largest cuts in carbon emissions (57 percent) come from the electricity sector (see Figure 7.2). Transportation delivers the next-largest reduction in global warming emissions, at 16 percent (or about 24 percent, if we remove cuts stemming from the 2007

Energy Independence and Security Act from the Reference case).[77]

Offsets from storing carbon in U.S. agricultural lands and forests, and international offsets mainly from avoided tropical deforestation, provide 11 percent of the cuts in carbon emissions. Reductions in emissions from direct fuel use in industry and buildings contribute 9 percent of the total drop. Cuts in non-CO$_2$ emissions deliver the remaining 7 percent.

7.2.2. National Consumer and Business Costs and Savings under the Blueprint

The Blueprint policies not only dramatically cut carbon emissions—they also save consumers and businesses money. Considering costs and savings together, consumers will see annual savings from the Blueprint of $464 billion in 2030 compared with the Reference case (see Table 7.1).

Americans will save $414 billion on their energy bills in 2030 (on their monthly electricity bills, and on

---

76 Apart from the Blueprint policies, the United States could spur another 10 percent reduction in global emissions by investing in forest protection in developing countries (Boucher 2008), and potentially an additional amount by investing in clean technology in those countries.

77 Cuts in heat-trapping emissions in the transportation sector include those from refining transportation fuels.

gasoline costs, for example), even though those bills include the cost of carbon allowances passed through to consumers and businesses in higher energy prices. These savings also take into account the costs of renewable electricity, carbon capture and storage, and renewable fuels that are passed on to consumers and businesses through slightly higher energy prices. Consumers and businesses save money because energy efficiency and conservation measures lower total energy use under the Blueprint.

Of course, these savings would not come free. In 2030, consumers and businesses would have to invest about $160 billion in more efficient appliances and vehicles, upgrades to buildings, improved industrial processes, and expanded transit. That would leave consumers and businesses with a net annual savings of $255 billion. What's more, revenues from auctioning carbon allowances would be recycled back into the economy, putting another $219 billion back into the pockets of both consumers and businesses.

The costs of implementing Blueprint policies include the nearly $8 billion that government and industry will have to invest in 2030 to cover R&D on energy efficiency and cleaner energy, plus tax credits and the implementation costs of pursuing other policies under the Blueprint (see Table 7.1).[78]

The costs and savings associated with Blueprint policies are spread throughout the economy. The net annual savings for consumers and businesses of $255 billion in 2030 include utility and gasoline bills that incorporate carbon costs, per-mile congestion fees, and the cost of energy-consuming products.

While our analysis recycled the revenues from auctioning carbon allowances back into the economy (half to consumers and half to businesses), that recycling could occur in ways that further lower costs or increase climate benefits. For example, government could use the funds to provide tax credits for purchases of more efficient vehicles and appliances, to increase renewable energy use, or to encourage land uses that store more carbon.

TABLE 7.1. Annual Blueprint Savings
(in billions of 2006 dollars)

	2015	2020	2025	2030
Energy Bill Savings	$ 39	$152	$271	$414
Energy Investment Costs	-40	-80	-123	-160
Net Consumer and Business Savings	$ -1	$ 72	$147	$255
Allowance Revenue Generated	+145	+181	+207	+219
Added Policy Implementation Costs	-9	-13	-8	-8
Blueprint Savings	$136B	$240B	$345B	$464B

Note: Values may not sum properly due to rounding.

Considering costs, savings, and recycling auction revenues, consumers and businesses will see annual savings from the Blueprint of $464 billion in 2030 (compared with the Reference case). These savings, of course, do not come for free. Consumers and businesses will need to invest in low-carbon energy technologies, efficiency, and conservation, and these investments quickly pay off with lower energy bills—using less electricity and fuel results in savings, even at slightly higher energy prices.

---

78 The cap-and-trade program will require moderate administrative costs. We were unable to quantify those costs explicitly in our analysis, but expect that they are too small to significantly influence our results.

FIGURE 7.3. The Source of Savings in 2030
(Blueprint case vs. Reference case)

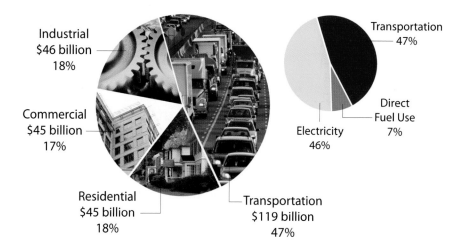

Industrial
$46 billion
18%

Commercial
$45 billion
17%

Residential
$45 billion
18%

Transportation
$119 billion
47%

Transportation
47%

Direct
Fuel Use
7%

Electricity
46%

Consumers and business-es see $255 billion in net annual savings in 2030 under the Blueprint (in 2006 dollars). Consumers and businesses in the transportation sector reap the largest share. Residential, commercial, and industrial consumers each gain just under 20 percent of the net sav-ings, with nearly 90 per-cent of that amount—or $118 billion—stemming from lower electricity costs.

7.2.3. Distributing the Costs and Savings under the Blueprint

The costs and savings associated with Blueprint poli-cies are spread throughout the economy. The net annual savings for consumers and businesses of $255 billion in 2030 include utility bills, gasoline bills, per-mile congestion fees, and the cost of energy-consuming products. However, those savings exclude any policy costs funded by general taxpayer revenues, the costs that utilities and fuel providers do not pass on to consumers, and the recycling of any revenues from auctions of carbon allowances.[79]

Based on end-use, transportation bears the largest portion of those costs, at 32 percent of the $160 bil-lion in energy investment costs in 2030, followed by the commercial sector at 25 percent. Industrial and residential consumers each carry slightly less than 20 percent of the energy investment costs in 2030.

Transportation users reap the largest share—40 percent—of the $414 billion in savings on energy bills in 2030 under the Blueprint. Residential, commercial, and industrial consumers each receive about 20 percent of the total savings, with savings on electricity bills accounting for more than 70 percent of the total.

Households and businesses that rely on the trans-portation sector see nearly half of the net annual sav-ings ($119 billion) in 2030 (see Figure 7.3). However, Blueprint policies ensure that consumers and busi-nesses throughout the economy save money on energy expenses. Lower electricity costs are responsible for $118 billion in net annual savings for industrial, com-mercial, and residential customers.

The net savings in 2030 are split almost evenly between businesses ($128 billion) and consumers ($126 billion), and are spread throughout all regions of the country (see Figure 7.4). The consumer sav-ings are also spread among the projected 140 million American households in 2030, cutting the annual household cost of energy and transportation by $900 that year compared with the Reference case.

7.2.4. National Economic Growth under the Blueprint

Under the Blueprint, gross domestic product (GDP) remains practically unchanged from the Reference case. In the latter, GDP grows from $11 trillion in 2005 to $20.2 trillion in 2030, an overall growth of 84 percent, and an average annual growth rate of 2.47 percent (in 2000 dollars).

Under the Blueprint, GDP grows from $11 trillion in 2005 to $19.9 trillion in 2030, an overall growth of 81 percent and an average annual growth rate of 2.41 percent. In the Blueprint case, GDP in 2030 is less than 1.5 percent below that in the Reference case—equivalent to only 10 months of economic growth over a 25-year period.[80] This shows that the nation can

---

79 A congestion fee of $0.006 per mile under the Blueprint would represent a charge to drivers for the cost of delays and pollution caused by congestion. The fees would be used to expand mass transit as an alternative to driving.

80 This means that, under the Blueprint, the economy reaches the same level of economic growth in October 2030 as the Reference case reaches in January 2030.

implement effective policies to tackle global warming without harming economic growth.

The Blueprint also shows practically the same employment trends as the Reference case. In fact, non-farm employment is slightly higher under the Blueprint than in the Reference case (170 million jobs versus 169.4 million in 2030).

Many other studies have also shown that the effects of such policies on the economy are small (see Keohane and Goldmark 2008 for a summary). And small differences are swamped by the uncertainty inherent in predicting GDP as far out as 2030. As Keohane and Goldmark point out, predictions from different models of GDP in 2030 can differ by as much as 10 percent.

Meanwhile the 2006 *Stern Review of the Economics of Climate Change* found that the costs of unchecked global warming could range from 5 to 20 percent of worldwide GDP, depending on the severity of climate change, by the end of this century (Stern 2006).

What's more, the NEMS model itself has serious limitations in its ability to account for the impact of Blueprint policies on GDP. For example, it is unable to fully consider the positive effects on GDP from investments in the energy and transportation sectors that enable consumers and businesses to save money on energy bills and spend it more productively. The model also does not include the effects on GDP of unchecked global warming in the Reference case.

7.2.5. Significant Reductions in Energy Use under the Blueprint

Under the Blueprint, total energy use is one-third (39 quadrillion Btu) lower than under the Reference case by 2030, and 23 percent below 2005 levels, because of a significant increase in energy efficiency in all sectors and with all fuels, as well as cuts in car and truck travel (see Table 7.2). Use of non-hydro renewable energy is 25 percent higher than in the Reference case by 2030,

FIGURE 7.4. Net Consumer and Business Savings
(by Census Region in 2030, in 2006 dollars)

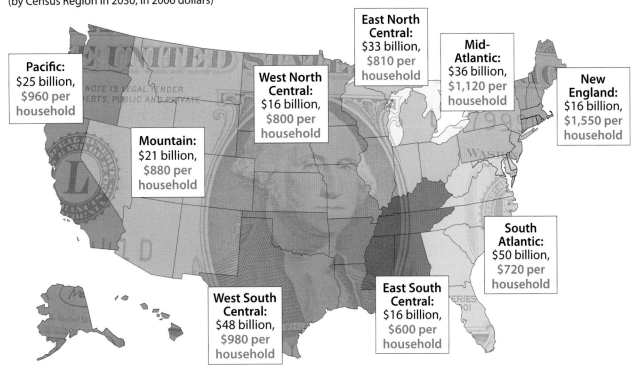

East North Central: $33 billion, $810 per household

Mid-Atlantic: $36 billion, $1,120 per household

Pacific: $25 billion, $960 per household

West North Central: $16 billion, $800 per household

New England: $16 billion, $1,550 per household

Mountain: $21 billion, $880 per household

South Atlantic: $50 billion, $720 per household

West South Central: $48 billion, $980 per household

East South Central: $16 billion, $600 per household

Net Annual Savings in 2030	Total	$255 billion
	Business	$128 billion
	Consumers	$126 billion
	Average Consumer	$900 per household

Note: Values may not sum properly because of rounding.

Consumers and businesses in every region of the country save billions of dollars under the Blueprint. Household numbers do not include business savings.

with that sector's share of total energy use rising to 21 percent by 2030, after accounting for improvements in energy efficiency.

Greater energy efficiency and use of renewable energy reduce coal use 85 percent by 2030 compared with the Reference case, with most of the cuts coming from the electricity sector. However, the Blueprint does show a modest increase in the use of advanced coal plants with carbon capture and storage before 2030 compared with the Reference case. That technology could play a more significant role if its cost declines faster than the Blueprint assumes, or if the nation does not pursue energy efficiency and renewable energy as aggressively.

Natural gas use is more than one-third lower in 2030 under the Blueprint compared with the Reference case, primarily because of energy efficiency improvements in industry and buildings, and more modest use of natural gas in power plants. Oil use is about 24 percent

lower in 2030, with most of the reduction occurring in transportation and industry. The use of nuclear and hydropower, which do not produce carbon emissions directly, is similar to that in the Reference case.

7.2.6. Curbing Our Oil Addiction under the Blueprint

The Blueprint reduces demand for oil and other petroleum products in 2030 by about 6 million barrels per day—or 30 percent—compared with 2005 (see Figure 7.5). That drops imports to less than 45 percent of U.S. demand for petroleum, compared with more than 60 percent in 2005.

Because the United States is the world's largest petroleum consumer, cutting U.S. demand by 30 percent helps hold oil prices to $80–$88 per barrel from 2020 to 2030—about $10 per barrel below Reference case projections. As a result of lower oil prices and reduced

TABLE 7.2. Comparison of U.S. Energy Use

(Blueprint case vs. Reference case, in quadrillion Btu)

Fuel	2005	2020 Reference Case	2020 Blueprint Case	2030 Reference Case	2030 Blueprint Case
Petroleum	40.1	37.9	33.4	38.1	28.8
Natural Gas	22.6	23.8	18.5	23.6	15.7
Coal	22.8	25.2	15.1	29.3	4.5
Nuclear Power	8.2	8.8	8.8	8.6	8.5
Hydropower	2.7	3.1	3.1	3.2	3.2
Other Renewables[a]	3.5	9.1	10.7	13.0	16.2
Other[b]	0.2	0.2	0.2	0.3	0.3
Total	100.1	108.0	89.8	115.9	77.2
Energy Savings					
vs. Reference case			17%		33%
vs. 2005			10%		23%

Notes:

a "Other renewables" include grid-connected electricity from landfill gas, biogenic municipal waste, biomass, wind, geothermal, solar photovoltaic and thermal sources, and non-electric energy from biofuels and active and passive solar systems. These values exclude imported electricity generated from renewable sources and non-marketed renewable energy.

b "Other" includes non-biogenic municipal waste and net electricity imports.

The Blueprint policies reduce projected U.S. energy use one-third by 2030, with the help of efficiency and conservation. Carbon-free electricity and low-carbon fuels together make up more than 33 percent of the remaining energy use; other renewable energy sources increase by 25 percent while nuclear and hydropower stay relatively flat.

demand, the United States spends about $550 million per day on oil imports in 2030—about $450 million less than in the Reference case.

Those savings could end up higher or lower depending on a variety of factors not included in the NEMS-UCS model. If political instability rises, or if world demand exceeds supply, the resulting spikes in oil prices could mean dramatically higher savings under the Blueprint. In fact, reduced demand for oil is an insurance policy against exactly that scenario. If OPEC nations respond by reducing supply to drive up prices and thus siphon off some of our savings, the U.S. economy will be much more resilient in the face of such tactics.

7.2.7. Economywide Growth in the Use of Bioenergy under the Blueprint

Use of bioenergy is projected to more than triple by 2030 under the Blueprint. That increase is driven first by the production and use of biofuels in the transportation sector, and second by the use of biomass to generate electricity. Bioenergy use in the industry and buildings sectors does not change significantly in the Blueprint case (see Figure 7.6).

While significant, the growth in biofuel use is almost the same as that in the Reference case, because most of that use stems from the national renewable fuel standard included in the 2007 Energy Independence and Security Act. By 2030, nearly two-thirds of the U.S. supply of bioenergy is used for biofuels.

As a result, total bioenergy use in the Blueprint case is only 16 percent higher by 2020, and 3 percent higher by 2030, than in the Reference case. Almost all of this increase occurs in the electricity sector, where biomass is burned with coal in existing coal plants over the near-term and mid-term, and in dedicated biomass power plants over the longer term, to help meet the national renewable electricity standard.

Increases in bioenergy use under the Blueprint are modest, because we assumed certain limits on the amount of cellulosic crops grown for energy use, to minimize direct and indirect carbon emissions. These limits, and significant increases in demand for biofuels, mean that nearly all cellulosic crops and agricultural residues are used for transportation fuels by 2030 in the Reference case.

Growth in the use of bioenergy is also limited by the assumption in UCS-NEMS that use of forest, mill, and urban residues is restricted to the electricity sector. Finally, the use of corn for biofuels dropped under the Blueprint because that use does not reduce carbon emissions compared with gasoline. These factors limit the

FIGURE 7.5. Demand for Petroleum Products

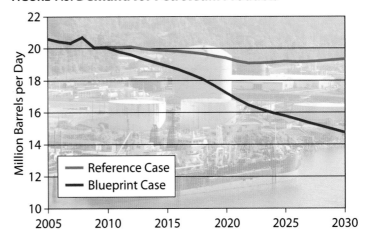

In 2030, the Blueprint cuts the use of petroleum products by 6 million barrels a day compared with 2005 levels. Because the United States is the world's largest petroleum consumer, cutting our demand this significantly could help lower oil prices.

Transitioning to a low-carbon energy system helps us kick our oil addiction and reduce our dependence on oil from politically troubled regions, such as the Middle East. By 2030, the Blueprint cuts the use of oil and other petroleum products by 6 million barrels per day—as much oil as the nation now imports from OPEC.

ability of the transportation sector to meet an even more stringent low-carbon fuel standard, which would have driven up the use of biofuels.

7.3. Detailed Results: The Blueprint Cap-and-Trade Program

The cap-and-trade program modeled as part of our Blueprint policies helps deliver the necessary level of cuts in global warming emissions. The next sections explore major findings related to key aspects of this program (described in Chapter 3).

FIGURE 7.6. Bioenergy Use
(Blueprint case vs. Reference case)

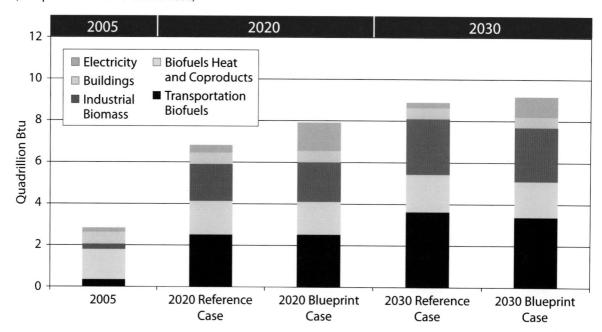

Note: Biofuels heat and coproducts represent the biomass energy that is left over from the process of turning that biomass into biofuels for the transportation sector. This energy ends up as useful heat, electricity, or animal feed.

U.S. bioenergy use is projected to more than triple by 2030 under the Blueprint, due mainly to increased use in the transportation and electricity sectors. While significant, the growth in biofuel use is similar to the Reference case because of the national renewable fuel standard included in the 2007 Energy Independence and Security Act. To minimize direct and indirect carbon emissions, the Blueprint assumed limits on the supply of energy crops; therefore, increases in bioenergy use are modest.

7.3.1. Prices of Carbon Allowances and the Resulting Revenues

The comprehensive policy approach in the Blueprint has a moderating effect on the prices of carbon allowances. These prices range from $18/ton in 2011 (the year the program starts) to $34/ton in 2020 to $70/ton in 2030 (all prices in 2006 dollars) (see Figure 7.9).

Those prices are well within the range that other analyses find, despite our stricter cap on economywide emissions. In addition, the Blueprint achieves much larger cuts in carbon emissions within the capped sectors because of the tighter limits that we set on offsets, and because of our more realistic assumptions about the cost-effectiveness of investments in energy efficiency and renewable energy technologies.

Under the Blueprint case, the revenues raised from auctioning 100 percent of allowances to emit carbon are significant, amounting to a cumulative total of $1.3 trillion by 2030 (in 2006 dollars, discounted at a 7 percent rate). Annual revenues range from $116 billion in 2011—the year the cap-and-trade program goes into effect—to $181 billion in 2020, and to $219 billion in 2030 (all figures in 2006 dollars).

We assumed that the government recycles these revenues directly back into the economy, so they represent a transfer payment rather than an actual cost of this policy. However, because of limitation in the UCS-NEMS model, we could not model a targeted way of recycling the revenues to specific purposes. We could only model recycling revenues in a general way to consumers and businesses.

The preferred approach would be to target revenues toward investments in energy efficiency, low-carbon technologies, and protection of tropical forests, as well as transition assistance to consumers, workers, and businesses to help them make the shift to a clean energy economy. Those uses would reduce carbon emissions and create additional economic benefits, such as savings on energy bills.

7.3.2. Banking and Withdrawing

We allowed companies subject to the cap-and-trade program to engage in unrestricted banking and withdrawing of carbon allowances, and assumed a final bank balance of zero in 2030. This is a flexibility mechanism that allows firms to choose a cost-effective path to cutting their emissions, and that reduces the volatility of the price of carbon allowances.

Our results show that the most cost-effective path to meeting the emissions cap is one in which firms overcomply with the cap requirements and accumulate

Under the Blueprint case, the revenues raised from auctioning 100 percent of allowances to emit carbon are significant, amounting to a cumulative total of $1.3 trillion by 2030 (in 2006 dollars, discounted at a 7 percent rate).

banked allowances until 2024. That result is typical in modeling cap-and-trade programs. For example, the Energy Information Agency's modeling of the Lieberman-Warner Climate Security Act of 2008 also showed a similar build-up of banked allowances (EIA 2008).[81]

We also find that firms run down the allowance bank to zero in 2030, a result driven by our assumption of a zero terminal bank balance.

As a result of this banking and withdrawing, the actual trajectory of carbon emissions under the model diverges from the trajectory set in the cap. For example, in 2020 U.S. heat-trapping emissions are 30 percent below 2005 levels—higher than the 26 percent required by the cap. In 2030 they are 44 percent below 2005 levels: lower than the 56 percent required by the cap. However, cumulative emissions—the critical metric—are the same under both trajectories (see Figure 7.7).

7.3.3. Prices of Carbon Offsets

High-quality carbon offsets—if limited—can play an important role in a cap-and-trade program (see Chapter 3). Our results show that in the early years of the Blueprint cap-and-trade program, many cost-effective opportunities for cutting emissions are available within the capped sectors, so firms do not need to use the full amount of offsets available to them.

The limit on domestic offsets that we modeled—amounting to 10 percent of the cap on global warming emissions—becomes binding starting in 2020. Until that year, the price of domestic offsets is the same as

---

81 In fact, in that case, because the modeling imposed a positive final bank balance of 5 billion metric tons, the results show banking through 2030.

FIGURE 7.7. Actual Emissions Compared with Cap Emissions

(Blueprint results vs. model input, 2000–2030)

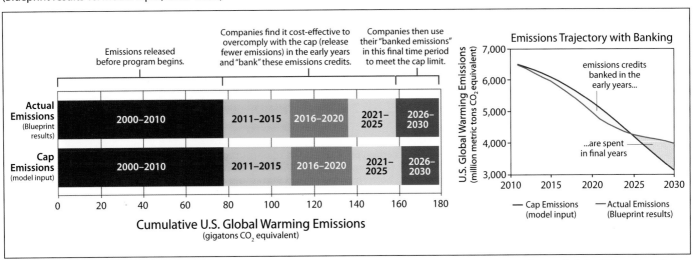

The bar graph shows two scenarios for cumulative emissions from 2000–2030. Although each scenario takes a slightly different path, the end point in 2030 is the same. The bottom bar in the graph corresponds to the cumulative emissions set under the cap, while the top is the actual cumulative emissions that emerged from our modeling results. From 2000–2010, before the start of the Blueprint cap-and-trade program, the cumulative emissions are the same in both cases. After 2010 the two trajectories diverge (actual cumulative emissions are lower than those required by the cap in the first three periods, and higher in the final period). What's important for the climate is that the United States stays within the emissions limits set by the cap-and-trade program.

the price of carbon allowances (for example, $18 per ton of carbon in 2011, and $34 per ton in 2020). After that point, the price of offsets drops below the price of carbon allowances, because offset providers now have to compete with each other to meet the limited demand. The price of domestic offsets drops to $26 per ton in 2025, and $18 per ton in 2030 (see Figure 7.8).

International offsets are available at a significantly lower price than that of carbon allowances and domestic offsets, based on our supply curve assumptions (see Appendix B online). The price of these offsets ranges from $10 per ton in 2011 to about $2 per ton in 2020, and to just more than $1/ton in 2030. Our limit on international offsets—amounting to 5 percent of the cap on emissions—becomes binding as soon as the cap-and-trade program begins.

Limits on offsets help ensure that the capped sectors make the needed long-term investments to reduce carbon emissions.

The domestic offsets we modeled are based on activities that increase carbon storage in agriculture and forests, such as changes in tillage practices, afforestation, and better forest management.[82] Because of scientific uncertainties in measuring emissions from these sectors, it is hard to cap the sectors directly, though they can be included in a cap-and-trade program as a (bounded) source of offsets. Forests and agriculture have a significant potential to contribute to U.S. global warming solutions, which specific (non-offset) policies targeting these sectors could encourage (see page 166).

Carbon storage in forests and soils is also subject to saturation or even reversal, so we cannot count on such offsets as a permanent solution to global warming. Eventually, forests and soils will stop absorbing carbon, and could even turn into net sources of carbon emissions.[83]

FIGURE 7.8. Prices of Carbon Allowances and Offsets under the Climate 2030 Blueprint

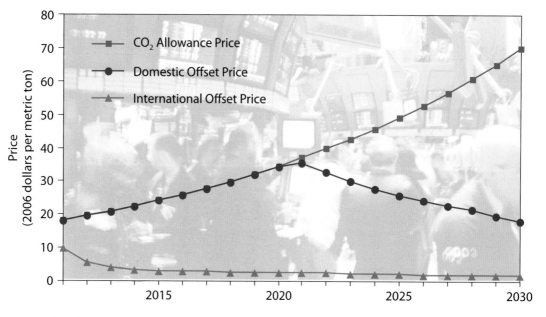

A cap-and-trade system generates a carbon price designed to encourage the capped sectors—such as electricity and oil refining—to lower their emissions and invest in clean technologies. The Blueprint also allowed a limited amount of offsets, both from the United States and other countries, as an alternative way for firms to comply with the cap. The prices of these offsets vary depending on their source, the relative cost of reducing emissions within the capped sectors, and whether the maximum limit on offsets is reached.

82 We used the supply curves for domestic offsets based on carbon sequestration in agriculture and forests embedded in the NEMS model. These, in turn, are based on information from the Environmental Protection Agency, derived from the FASOMGHG model (see Section 7.7.2). Although we have concerns about the criteria used to construct these supply curves, we were unable to find enough robust data to construct different ones.

83 Without policy intervention, many forests are poised to release carbon now, given droughts, fires, and pest outbreaks associated with global warming, as well as poor management practices.

FIGURE 7.9. Carbon Dioxide Emissions from Power Plants

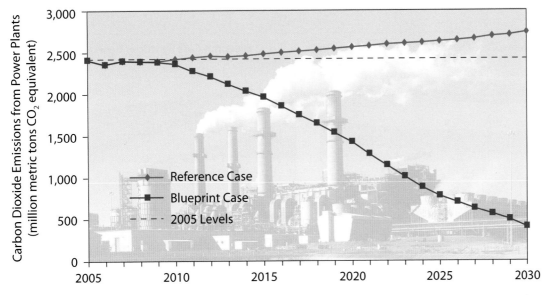

Carbon emissions from power plants grow nearly 14 percent by 2030 under the Reference case, as coal use increases to help meet projected growth in electricity demand. Under the Blueprint, however, power plant carbon emissions are 84 percent below 2005 levels by 2030. Sulfur dioxide, nitrogen oxide, and mercury emissions from power plants are also significantly lower under the Blueprint, which, by improving air and water quality, would provide public health benefits.

The international offsets we modeled are based primarily on reduced emissions stemming from avoided tropical deforestation.[84]

7.4. Detailed Results: The Electricity Sector

7.4.1. Reference Case: Carbon Emissions from Power Plants Grow

Under the Reference case, carbon emissions from power plants continue to rise over time, as fossil fuel use increases to help meet growth in electricity demand (see Figure 7.9). By 2030, CO_2 emissions from power plants grow by nearly 14 percent over 2005 levels. The Reference case projects that U.S. electricity use will grow 25 percent from 2005 to 2030, because technologies and practices to encourage energy efficiency will be underused.

Nearly all of the increase in carbon emissions from power plants in the Reference case is due to expanded use of coal to produce electricity, which remains the dominant fuel for that use. Coal-based electricity grows 29 percent by 2030, as the nation builds 61 gigawatts of new capacity—the equivalent of more than 100 new 600-megawatt coal plants. That is considerably lower than the EIA's projection in 2008 that the nation will

have 104 gigawatts of new coal capacity by 2030. However, it is about one-third higher than the agency's projection in 2009 that the nation will have 46 gigawatts of new capacity by 2030 (EIA 2008a; 2009).

Electricity produced from natural gas, nuclear, and combined-heat-and-power (CHP) plants all remain relatively unchanged in the Reference case (see Figure 7.10). While the nation adds 87 gigawatts of new natural gas capacity by 2030, most of these plants displace older, less efficient natural gas plants, or produce electricity only during periods of high demand. And while loan guarantees and tax credits available under current law spur construction of 4.4 gigawatts of new nuclear capacity (four new plants), this replaces a similar amount of nuclear capacity that will go out of service when the 20-year license extensions of today's plants expire.

Electricity from renewable resources, including wind, solar, geothermal, and biomass, expands from 3 percent of total demand in 2008 to about 10 percent in 2030 in the Reference case. That increase in market share is due largely to state renewable electricity standards, federal tax credits, and an increase in combined heat and power from new biofuel plants built under

---

84 For 2011 to 2015, we used the international offsets supply curve developed by the EIA for NEMS, which is based on data from the EPA (see footnote 9). For 2015 to 2030, we used a supply curve based solely on offsets from avoided tropical deforestation, developed by UCS analysts (see Appendix B online for more details).

FIGURE 7.10. Sources of Electricity

(Reference case vs. Blueprint case)

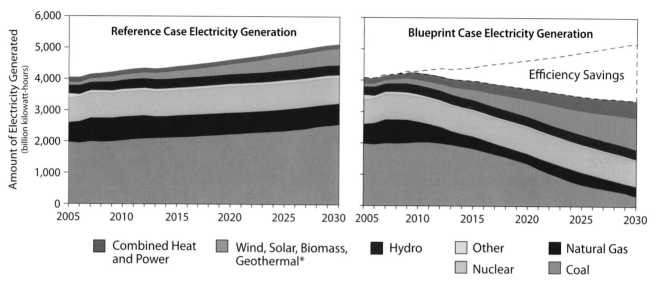

* Landfill gas and incremental hydro are also included in this category.

The Blueprint reduces electricity demand and diversifies our electricity mix. If we continue on our current path (as represented by the Reference case), electricity use continues to grow and this increased demand is met primarily by carbon-intensive coal-fired power plants. The Blueprint, conversely, reduces electricity demand 35 percent in 2030 through aggressive energy efficiency measures, while generation from efficient combined-heat-and-power plants more than triples over current levels, and renewable electricity expands to 40 percent of the nation's total electricity use.

the federal renewable fuel standard (see Section 7.6). Most of the increase in renewable electricity comes from wind and bioenergy, followed by geothermal and distributed solar photovoltaics (PV). Hydroelectric power remains relatively unchanged, providing 6 percent of U.S. electricity by 2030.

7.4.2. Blueprint Case: Dramatic Cuts in Power Plant Emissions

Under the Blueprint, the electricity sector makes the biggest contribution to reducing U.S. global warming emissions, providing 57 percent of all cuts in 2030, compared with the Reference case. Carbon emissions from power plants are 41 percent below 2005 levels by 2020, and 84 percent below 2005 levels by 2030. Sulfur dioxide (SO_2), nitrogen oxide (NOx), and mercury emissions from power plants are also significantly lower under the Blueprint, which would improve air and water quality and thus provide important public health benefits.

Most of the cuts in emissions in the electricity sector come from replacing coal plants with efficiency, combined heat and power, and renewable energy under the Blueprint (see Figure 7.10). By 2030, energy efficiency measures—such as advanced buildings and

industrial processes, and high-efficiency appliances, lighting, and motors—reduce demand for electricity 35 percent below the Reference case. CHP based on natural gas in the industrial and commercial sectors is nearly 3.5 times higher than today's levels, providing 16 percent of U.S. electricity by 2030. Largely because of the national renewable electricity standard, wind, solar, geothermal, and bioenergy provide 40 percent of the nation's electricity use by 2030, after accounting for the drop in demand stemming from energy efficiency and CHP.

The increase in energy efficiency, CHP, and renewable energy spurred by the Blueprint policies—combined with a cap-and-trade program that requires owners of fossil fuel plants to buy allowances to emit carbon—significantly reduces coal-based power by 2030. Owners of many existing coal plants opt to co-fire biomass with coal, to reduce their emissions in the early years. A few existing coal plants are also replaced with advanced coal plants with carbon capture and storage. If the cost of this technology declines faster than the Blueprint assumes—or if the nation does not deploy energy efficiency measures and renewable energy as extensively—coal generation would not decline as much as Figure 7.10

shows, and coal-burning power plants would emit fewer carbon emissions.

Coal use in power plants declines from more than 1 billion tons in 2005 to 137 million tons in 2030 under the Blueprint, compared with an increase to more than 1.2 billion tons in the Reference case. The Blueprint displaces a cumulative total of more than 11 billion tons of coal use in power plants through 2030, producing important environmental and public health benefits (see Box 7.3).

The Blueprint policies do not spur a widespread switch to natural gas from coal to produce electricity, as other studies have projected. In fact, the amount of electricity from stand-aloxne power plants burning natural gas is nearly one-third lower under the Blueprint than in the Reference case by 2030. However, an increase in electricity production from CHP based on natural gas in the commercial and industrial sectors more than offsets this drop. Electricity producers use less natural gas under the Blueprint because CHP plants use more waste heat than stand-alone power plants, and are therefore much more efficient.

Under the Blueprint, new advanced (integrated gasification combined-cycle) coal plants with carbon capture and storage (CCS), and advanced nuclear power plants, play a very limited role before 2030, as these technologies are not economically competitive with other options during that time frame. The Blueprint includes 7 gigawatts of capacity from new advanced coal plants with carbon capture and storage, including 4.8 gigawatts from eight large-scale plants built as a result of our recommended CCS demonstration program. The model also adds nearly 3 gigawatts of new natural gas capacity with CCS by 2030.

The model does not add any advanced nuclear plants by 2030 beyond the 4.4 gigawatts of new capacity added in the Reference case. However, almost all existing nuclear plants continue to operate, because they do not emit carbon, and their owners therefore do not have to purchase carbon allowances.

7.4.3. Blueprint Case: Renewable Energy Diversifies the Electricity Mix

Because of the national renewable electricity standard (RES) in the Blueprint, power producers generate almost twice as much electricity from wind, solar, geothermal, and biomass as in the Reference case by 2030, using a more diverse mix of technologies (see Figure 7.11). Wind power makes the largest contribution, providing nearly half of all renewable electricity by 2030. While most of this wind power is land-based,

One in four children in Harlem suffer from asthma, and already poor air quality will likely get worse as temperatures rise. The Climate 2030 Blueprint dramatically improves air and water quality and protects public health by reducing carbon emissions from power plants, as well as emissions that cause acid rain, smog, and mercury pollution.

FIGURE 7.11. Blueprint Renewable Electricity Mix (2030)

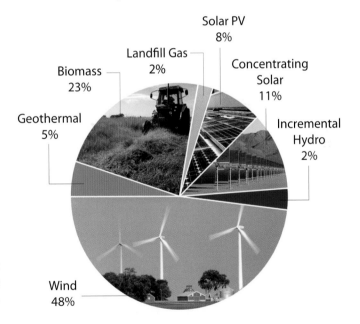

Solar PV 8%
Landfill Gas 2%
Concentrating Solar 11%
Biomass 23%
Geothermal 5%
Incremental Hydro 2%
Wind 48%

the model projects that developers will build a small amount of offshore wind near the end of the period.

Biomass also makes a significant contribution. In the near-term, most biomass is co-fired with coal in existing coal plants. With a price on carbon emissions under a cap-and-trade program, cofiring becomes an attractive strategy, enabling owners of coal plants to meet near-term targets for cutting emissions. After 2020, cofiring declines as owners retire coal plants, and most biomass is used to produce biofuels in plants with

CHP, and to produce electricity in advanced biomass gasification plants.

Solar photovoltaics (PV) and concentrating solar thermal plants that can store electricity also expand significantly under the national RES, combined with the national solar investment tax credit, solar requirements in state RES policies, and other state policies in the Reference case. Together these policies spur deployment of solar, owing partly to greater economies of scale in manufacturing, constructing, operating, and maintaining it, making it competitive with other renewable energy technologies over time.

By 2030 the amount of variable power from wind and PV rises to 20 percent of the U.S. electricity supply,

---

BOX 7.2.

Public Health and Environmental Benefits of Reduced Coal Use

Cumulatively, the Blueprint displaces the need for more than 11 billion tons of coal for power plants by 2030 compared with the Reference case. Displacing that much coal would provide environmental and public health benefits roughly equivalent to:

- 280,000 premature deaths avoided
- 140,000 hospital admissions avoided
- 440,000 heart attacks avoided

- 6,400,000 asthma attacks avoided
- 1.1 million pounds of toxic mercury pollution avoided
- 12,600 square miles of surface mining avoided
- 130 square miles of Appalachian ridgeline saved
- 350 square miles of mountaintop removal mining avoided
- 5.6 billion gallons of mining slurry ponds avoided, equal to the volume of 520 Exxon Valdez spills

In December 2008, more than a billion gallons of fly ash sludge—a by-product of coal-fired electricity generation—breached a holding pond at a power plant near Harriman, TN. The sludge poured into the Emory River, flooding nearby homes and fields. Reducing reliance on coal plants over time by increasing efficiency and renewable energy yields the Blueprint's biggest carbon reductions, as well as significant environmental and safety benefits.

See Appendix D online for assumptions and sources.

after accounting for the drop in demand from energy efficiency and CHP. Several studies by U.S. and European utilities have found that wind power can provide as much as 25 percent of annual electricity needs without undermining reliability, and that the cost to integrate that power into the electricity grid would be modest (Holttinen et al. 2007). A 2008 study from the U.S. Department of Energy (DOE) found that wind power could provide 20 percent of U.S. electricity by 2030, with no adverse impacts on the reliability of the electricity supply or any need to store power. That study found that wind power would cost the average household an extra 50 cents per month—not including federal incentives or any value for reducing carbon emissions (EERE 2008).

While electricity from geothermal power plants more than triples from today's levels by 2030, virtually all of this increase occurs in the Reference case in response to existing state and federal policies. The vast majority of this development is new hydrothermal projects in the western United States. While geothermal makes an important contribution to electricity needs in that region, it makes a fairly modest contribution at the national level, because very little development of enhanced geothermal systems (EGS) is projected to occur before 2030. However, EGS has the potential to make a large contribution to reducing emissions after 2030, or if its cost declines faster than our analysis assumes. Electricity produced from landfill gas and incremental hydro (reflecting greater efficiency and expansion at existing plants) also makes a modest contribution to the national RES, given limited potential for these resources.

7.4.4. Blueprint Case: Significant Savings on Electricity Bills

Under the Blueprint, consumers and businesses in all sectors of the economy would see lower electricity bills compared with the Reference case. Annual savings would top $82 billion in 2020, and grow to $175 billion in 2030. Those savings would be offset somewhat by the cost of investments in energy efficiency and combined heat and power.[85] However, electricity customers would still see a net annual savings of $49 billion in 2020, and $118 billion in 2030. And

the average household would see net annual savings of more than $110 in 2020, rising to $250 in 2030.

Under the Blueprint, average electricity prices are nearly 8 percent higher than in the Reference case by 2020, and 17 percent higher by 2030. Those price increases mainly reflect the cost of carbon allowances, which raise the cost of burning coal and natural gas to produce electricity, and the cost of replacing existing coal and natural gas plants with new renewable energy

> A 2008 study from the U.S. Department of Energy (DOE) found that wind power could provide 20 percent of U.S. electricity by 2030, with no adverse impacts on the reliability of the electricity supply or any need to store power.

facilities and coal CCS projects from our demonstration program. However, while electricity prices are slightly higher under the Blueprint, consumers still save money on their energy bills because of reductions in electricity use from energy efficiency measures.

7.5. Detailed Results: Industry and Buildings
7.5.1. Reference Case: Emissions Rise as Homes and Businesses Use More Energy

Buildings now account for 40 percent of U.S. primary energy use, while industry accounts for nearly one-third.[86] Under the Reference case, primary energy use rises by 22 percent in residential and commercial buildings, and 10 percent in industry, from 2005 to 2030, because measures to boost energy efficiency are underused.

Almost all of the increase in primary energy use in buildings results from more electricity use, noted above. The use of natural gas increases slightly and the use of oil declines slightly over time, primarily because oil prices rise faster than natural gas prices. In industry,

---

85   Electricity prices and consumer bills already reflect the additional costs of investments in new renewable energy, fossil fuel, and nuclear facilities for generating electricity.

86   Primary energy use includes direct fuel use by homes and businesses for heating, cooling, and other needs, as well as indirect fuel use for generating electricity, which is allocated to each sector based on its share of electricity demand.

BOX 7.3.

The Perfect Storm of Climate, Energy, and Water

The reductions in electricity from coal and other fossil fuels resulting from greater energy efficiency and reliance on renewable energy will save significant amounts of water (see Table 7.3). In 2030 the nation would see a net drop in water use of more than 1 trillion gallons—equivalent to today's annual water use by 32 million people, or nearly three times the volume of Lake Erie. Cumulative water savings between 2010 and 2030 would reach nearly 12 trillion gallons. Reductions in water use at coal and other fossil fuel plants would offset modest increases in water use at bioenergy and concentrating solar thermal plants. Those water savings will be important in regions such as the West and the Southeast, where water shortages and drought will become more severe with global warming.

TABLE 7.3. Water Savings in Electricity Generation (2030)[a]

(Blueprint case vs. Reference case)

Type of Power Plant	Billion Gallons of Water
	Reduced Water Use
Coal	1,210
Natural Gas	81
Oil	29
Subtotal	1,320
	Increased Water Use
Bioenergy[b]	43
Concentrating Solar Thermal[c]	7–64
Subtotal	50–107
Net Water Savings under Blueprint	1,183–1,241

Notes:

a Reductions in water consumption are based on a drop in electricity produced from fossil fuels and an increase in renewable generation under the Blueprint. See Appendix D online for assumptions and sources.

b This includes only water used at the power plant. Biomass residues require no additional water. The amount of water used to grow energy crops (mainly switchgrass) is negligible, as we assumed that energy crops would grow on land that does not need irrigation.

c The range represents the use of dry cooling versus wet cooling. Dry cooling is more common in the Southwest, where the vast majority of concentrating solar plants will be located.

In a warming world, more precipitation in mountain regions such as California's Sierra Nevada will fall as rain instead of snow; the snow that does fall will melt earlier, drastically reducing the spring snowpack that provides water for people and agriculture. Water shortages are likely to become acute and widespread, especially in the western and southeastern United States.

This Oregon, OH, coal plant sits on the shores of Lake Erie, situated there to take advantage of lake water for cooling. By reducing the amount of fossil-fuel-generated electricity, the Blueprint would save more than 1 trillion gallons of water in 2030—equivalent to nearly three times the volume of Lake Erie or the amount of water used today by 32 million people.

the increase in energy use is due mostly to an increase in the production of liquid coal and biofuels.

Growing use of fossil fuels in these sectors, combined with more electricity use in buildings, means that carbon emissions from buildings rise 17 percent above 2005 levels by 2030, while those from industry rise 7 percent.

7.5.2. Blueprint Case: Efficiency Greatly Reduces Energy Use

Under the Blueprint, industry and buildings are responsible for 9 percent of all reductions in global warming emissions from direct fuel use. Efforts by industry and building owners to increase efficiency, CHP, and renewable energy also drive a significant portion of the reductions in emissions from the electricity sector noted above. If we assign those cuts to industry and buildings, their share of total reductions in global warming emissions would rise to 18 percent for industry and 48 percent for buildings.

Energy use in industry and buildings is dramatically lower under the Blueprint because the suite of standards, incentives, and other policies spurs greater energy efficiency and use of combined heat and power. Primary energy use in industry is 37 percent lower by 2030 under the Blueprint compared with the Reference case. That includes a 69 percent reduction in fuel used to generate electricity, a 63 percent reduction in coal use, a 23 percent reduction in oil use, and a 19 percent reduction in natural gas use.

Primary energy use in buildings is 40 percent lower by 2030 compared with the Reference case. That includes a 40 percent reduction in the use of fuel to generate electricity, a 31 percent reduction in the use of natural gas, and a 35 percent reduction in the use of oil.

The reduction in natural gas use from energy efficiency measures is offset somewhat by an increase in natural gas use for CHP in the commercial and industrial sectors. However, the increase in CHP from natural gas reduces the need to purchase electricity from centralized power plants. Such plants are considerably less efficient because they typically do not use their waste heat, and because electricity is lost when transported from the power plant to the user. Therefore replacing these plants with CHP based on natural gas spurs a net drop in the use of natural gas and in carbon emissions.

7.5.3. Blueprint Case: Lower Energy Bills for Homes and Businesses

Under the Blueprint, the industry and buildings sectors would see lower energy bills compared with the

Primary energy use in buildings is 40 percent lower by 2030 (compared with the Reference case) due to the Blueprint's strong suite of policies that increase energy efficiency and combined-heat-and-power systems in our homes, businesses, and industries. The Solaire Apartments in New York City's Battery Park are a case in point. This LEED Gold-certified complex achieved aggressive goals for reducing energy and water use as well as peak electricity demand by using daylighting, "Low-E" windows, programmable thermostats, and Energy Star appliances in each unit.

Reference case. In 2030, total annual savings on energy (including the use of electricity, natural gas, oil, and coal) would reach nearly $243 billion. That figure includes $77 billion in the residential sector (or $550 per household), $87 billion in the commercial sector, and $79 billion in the industrial sector.

The cost of investing in energy efficiency measures would offset these savings somewhat. However, net annual savings would reach $136 billion in 2030, including $45 billion in the residential sector ($320 per

(continued on page 146)

BOX 7.4.

SUCCESS STORY
Some Good News in Hard Times

Training opportunities in the renewable energy industry are expanding rapidly—such as at this wind turbine blade assembly plant in North Dakota—and many skills used in conventional industries are easily transferable to clean energy jobs.

General Electric's vice president of renewable energy made headlines in 2008 when he promised to hire every graduate of Mesalands Community College's wind power program for the next three years (*NMBW* 2008). Although a guaranteed job offer isn't standard for people training for careers in renewable energy, expected job growth in the industry is good news.

The solar industry estimates that more than 15,000 jobs were created in 2007 and 2008 (SEIA 2009), and the wind industry boasts more than 35,000 new direct and indirect jobs created in 2008 (AWEA 2009c). U.S. manufacturing of wind turbines and their components has also greatly expanded, with more than 70 new facilities opening, growing, or announced in 2007 and 2008. The industry estimates that these new facilities will create 13,000 high-paying jobs, and increase the share of domestically made components from about 30 percent in 2005 to 50 percent in 2008 (AWEA 2009b).

Although job numbers for the entire renewable energy industry are difficult to find, data from individual sectors such as solar and wind attest to demand for skilled labor. With the Obama administration's promise of green jobs spurred by federal policies designed to

bring more renewables online, "clean-tech" careers will continue to grow—welcome news given that the U.S. economy shed 1.2 million jobs in the first 10 months of 2008 (BLS 2008).

Several studies have found that renewable energy projects can create more jobs than using coal and natural gas to generate electricity. For example, a recent Union of Concerned Scientists study found that a national renewable electricity standard of 25 percent by 2025 would create nearly 300,000 new jobs in the United States—or three times more jobs than producing the same amount of electricity from coal and natural gas (UCS 2009). The U.S. Department of Energy recently reported that the wind industry will create more than 500,000 new U.S. jobs if 20 percent of the nation's power comes from wind by 2030 (EERE 2008). A third study showed that manufacturing the components for wind, hydro, geothermal, and solar systems could create more than 381,000 U.S. jobs (Sterzinger and Svrcek 2005).

As demand for workers has grown, so too has the number of schools devoted to training people for jobs in

> **The solar industry estimates that it created more than 15,000 jobs in 2007 and 2008; the wind industry boasts that it created more than 35,000 new direct and indirect jobs in 2008.**

renewable energy. Besides New Mexico's Mesalands—whose students are guaranteed employment with GE—Highland Community College in Illinois and Laramie County Community College in Wyoming introduced wind

Flexible, thin-film solar photovoltaic cells, such as those produced at California-based Nanosolar Inc., allow more homes and buildings to harness sunlight to generate electricity.

Renewable energy and energy efficiency jobs can be found in every region of the country. Green For All (*www.greenforall.org*) promotes clean energy job training in communities with chronically high unemployment rates to ensure that a clean environment and a strong economy go hand in hand.

technician programs in 2008. Colorado's Solar Energy International instructs 2,500 students each year in alternative energy systems. And enrollment in engineering for alternative energy at Lansing Community College in Michigan has jumped from 20 to 158 students since 2005, according to program staff (Glasscoe 2009).

Although training programs for jobs in renewable energy have expanded, many skills used in conventional industries such as manufacturing are transferable with no additional training. After a small Iowa town lost more than 100 jobs with the closing of a local plant making hydraulics, for example, most found new employment with wind turbine manufacturer Acciona after it converted the plant to build turbines (Goodman 2008).

Near Saginaw, MI, Hemlock Semiconductor provides the raw materials for electronic devices such as cell phones and, increasingly, solar panels. When completed in 2010, Hemlock's expansion to serve its growing solar business means the company will add 250 full-time jobs and 800 temporary construction jobs in a state that shed more than 400,000 jobs from 2000 to 2007 (Fulton and Cary 2008; Hemlock Semiconductor Corp. 2007).

Jobs in renewable energy are also geographically diverse: from staffing geothermal energy systems in Alaska to manufacturing biomass pellets in Florida. And while renewable energy can provide an important source of income and jobs for rural areas where many projects are located, they can create new manufacturing, construction, operation, and maintenance jobs in urban areas as well. The national group Green for All, for instance, works with cities such as Richmond, CA, to offer free training programs in trade skills for renewable energy (Apollo Alliance and Green for All 2008; Lee 2008).

Expanding the nation's use of renewable energy is essential to reducing our carbon emissions. In difficult economic times, the job growth spurred by clean, homegrown energy offers even more reason to ramp up its development.

By 2030, net annual electricity savings would reach $118 billion for industrial, commercial, and residential consumers under the Blueprint. Minnesota's Great River Energy is a generation and transmission cooperative providing wholesale electric service to other co-ops. The company's LEED Platinum-certified headquarters combines energy efficiency with on-site renewable energy and modest amounts of grid-supplied clean power to reduce fossil fuel use by 75 percent, cut CO_2 emissions by 60 percent, and save nearly 50 percent on energy costs compared with minimum building and equipment standards.

household), $45 billion in the commercial sector, and $46 billion in the industrial sector.

7.6. Detailed Results: Transportation

7.6.1. Reference Case: Carbon Emissions Climb Despite EISA

Our Reference case shows that carbon emissions from the transportation sector will grow by 12 percent between 2005 and 2030 (see Figure 7.12). Emissions are almost flat during the first two decades, growing only 2 percent between 2005 and 2022. This is due largely to the 2007 Energy Independence and Security Act (EISA), which requires cuts in carbon emissions from the production of most biofuels through 2022, and better fuel economy for cars and light-duty trucks through 2020. Once these policies stall out, however, carbon emissions in the transportation sector begin to grow at near historic rates.

Fuel economy for light-duty vehicles remained essentially stagnant from 1985 to 2005, as the auto industry successfully fought back attempts to require improvements in that metric. EISA pushes the fuel

economy of cars and light-duty trucks from about 25 miles per gallon in 2005 to more than 35 mpg in 2030. However, that falls short of the doubling in the fuel economy of new vehicles that existing technology could deliver. EISA also does not set specific efficiency targets for any other part of the transportation sector.[87]

EISA will help increase the share of low-carbon biofuels from just 0.1 percent of transportation fuel in 2005 to 9 percent by 2030. This significant increase highlights the importance of the requirement under the renewable fuel standard in EISA that limits carbon emissions from most biofuels. That requirement will bring low-carbon cellulose-based biofuels to scale, where they could become cost-competitive with petroleum.

Without EISA, we estimate that carbon emissions from the transportation sector would increase by about 30 percent instead of just 12 percent by 2030.[88] However, a transportation sector that simply runs in place on carbon emissions for a little over 10 years and then begins to increase again is not good enough. To actually cut carbon emissions compared with those in 2005, we need to go beyond the first step taken by EISA.

Even though the Reference case includes EISA, Blueprint policies will have to overcome the fact that emissions from cars and light-duty trucks drop only slightly in 2030 in the Reference case, while those from freight trucks and buses grow by nearly 40 percent, and those from airplanes rise by more than two-thirds (see Figure 7.13).

7.6.2. Blueprint Case: Driving Significant Cuts in Carbon Emissions

Blueprint policies for the transportation sector represent the essential next step after EISA. These aggressive but achievable policies address the three legs of the transportation stool: vehicles, fuels, and miles traveled for cars and light-, medium-, and heavy-duty trucks. When we add our Blueprint policies to the progress that occurs under EISA, the transportation sector can deliver a 19 percent reduction in carbon emissions in 2030 compared with 2005 (see Figure 7.12 and Figure 7.14).

That 19 percent drop stems from a cut in carbon emissions from transportation of more than 660 million metric tons in 2030—about 16 percent of the

---

87 EISA does require fuel-economy standards for medium- and heavy-duty trucks, but sets no specific minimum. EISA does not address fuel economy standards for planes, trains, off-road vehicles, or ships.

88 We estimate that EISA would reduce projected emissions by 350–450 MMTCO$_2$ in 2030. If automakers met the minimum EISA requirement of 35 mpg by 2020, emissions from transportation would decline by 250–300 MMTCO$_2$ in 2030. Wider use of low-carbon fuel under EISA is projected to save 100–150 MMTCO$_2$ in 2030.

FIGURE 7.12. Transportation Carbon Emissions
(Reference case vs. Blueprint case)

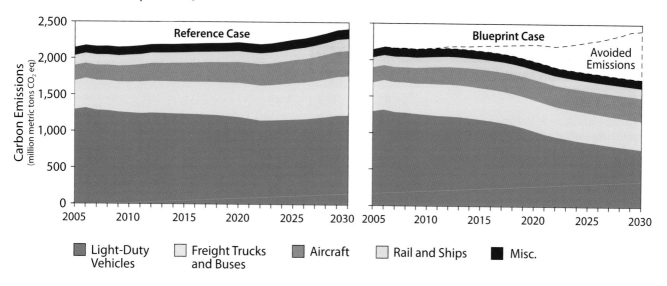

Light-Duty Vehicles   Freight Trucks and Buses   Aircraft   Rail and Ships   Misc.

Emissions are almost flat during the next two decades, due largely to the 2007 Energy Independence and Security Act (EISA) that will cut carbon emissions using biofuels and better fuel economy. However, a transportation sector that simply runs in place on carbon emissions for a while and then increases again is not good enough. To actually cut carbon emissions compared with those in 2005, we need to do more.

FIGURE 7.13. Changes in Transportation Carbon Emissions
(Reference case)

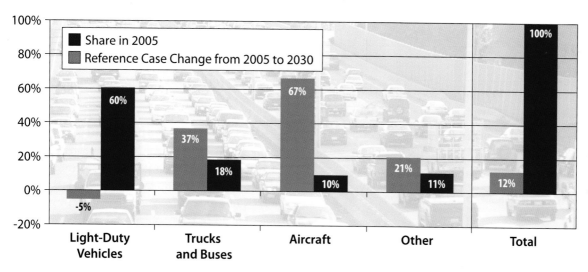

Under the Reference case, which includes EISA, carbon emissions from cars and light-duty trucks drop only slightly in 2030, while those from freight trucks and buses grow nearly 40 percent and those from aircraft increase by two-thirds.

carbon emissions saved that year. If we include the cuts in emissions spurred by EISA, transportation's contribution to total reductions in 2030 rises to more than 1 billion metric tons—or 24 percent of the reductions that year.

7.6.3. Blueprint Case: Greater U.S. Energy Security

The Blueprint delivers more than cuts in carbon emissions: it also improves energy security by reducing U.S. demand for oil, making our economy less vulnerable

to oil price shocks. While EISA keeps the amount of oil used for transportation from growing under the Reference case, Blueprint policies cut transportation's demand for oil and other petroleum products by 23 percent in 2030 compared with 2005.

Transportation provides more than half (53 percent) of the cuts in petroleum use achieved under the Blueprint. That represents savings of 3.2 million barrels per day in 2030, on top of the more than 3 million barrels of oil saved through EISA alone.

TABLE 7.4. Annual Consumer and Business Savings from Transportation

(in billions of 2006 dollars)

TRANSPORTATION SAVINGS	2020	2030
Fuel Cost Savings	$ 41	$ 172
Expense of Cleaner Vehicles and Reducing Vehicle Miles Traveled	-18	-53
Net Consumer and Business Savings	$ 23B	$ 119B

Savings at the pump from cleaner cars and trucks, better transportation choices, and low-carbon biofuels more than offset the costs of investing in these technologies. When these savings are spread out, the average U.S. household will save $580 per year by 2030 on annual transportation costs (versus the Reference case) and businesses that rely on transportation will save $38 billion.

FIGURE 7.14. Transportation Carbon Emissions (2030)

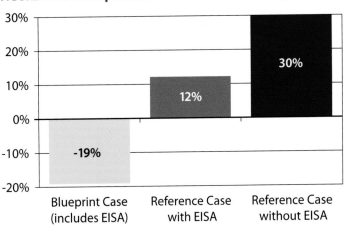

In 2030 the Blueprint delivers a 19 percent reduction in carbon emissions from the transportation sector compared with 2005. Under the Reference case, carbon emissions grow by 12 percent from 2005 to 2030. Had the 2007 Energy Independence and Security Act (EISA) not passed in 2007, transportation carbon emissions would have risen about 30 percent by 2030.

7.6.4. Blueprint Case: Saving Consumers and Businesses Money

By cutting fuel use through energy efficiency and reduced travel, and shifting transportation to cost-competitive, low-carbon fuels, Blueprint policies actually save consumers and businesses money while delivering cuts in carbon emissions. Through 2030, consumers and businesses will see their net annual expenditures on transportation—including the costs of fuel and vehicles—drop by about $120 billion compared with the Reference case (see Table 7.4). Of that savings, more than two-thirds—$81 billion—will end up in the hands of consumers in 2030, while businesses that rely on transportation will save $37 billion.

In other words, annual savings from the Blueprint transportation policies in 2030 not only cover the $53 billion cost of more efficient vehicles, better fuels, and new transportation alternatives, but also reward consumers and businesses who help cut carbon emissions.

Under the Blueprint, the average household will save $580 per year by 2030 on annual transportation costs versus the Reference case—and the average new vehicle will already get 35 miles per gallon in that baseline case. What's more, that figure excludes the potential for every vehicle owner to save as much as $150 per year on insurance costs owing to reduced driving (Bordoff and Noel 2008).

In earlier years, Blueprint policies ask consumers to invest in new technologies, such as better engines and transmissions and GPS monitoring systems, which will also enable pay-as-you-drive insurance. However, those technologies more than pay for themselves.[89] And total household savings would be even larger if they included the effects of recycling revenue from allowance auctions. For example, government could return such revenues as tax credits to consumers who purchase cleaner vehicles and fuels, or through other policies.

7.6.5. Blueprint Case: Keeping Gasoline Prices Down

Despite carbon allowances that cost as much as $70 per ton, gasoline prices rise only $0.10 per gallon under the Blueprint through 2020 compared with the Reference case, and no more than $0.24 from 2020 to 2030. Consumers pay up to $0.55 per gallon to cover the costs of carbon allowances passed on by oil companies. However, wholesale gasoline prices are $0.15-$0.40 per gallon below those of the Reference case from 2020 to 2030, owing to lower U.S. demand for oil and gasoline.

Those results contrast sharply with claims that a cap-and-trade program will significantly drive up fuel prices. The results point instead to changes in gasoline prices that are similar to or even lower than price spikes that have occurred within a few months or even weeks during the last few years. Including transportation in a cap-and-trade program will not significantly

---

89 These values assume that consumers pay the full incremental price of technologies the first year. Typical consumers will lease or obtain a loan on their vehicle, which would lower their costs in the early years.

Reducing emissions from the transportation sector will require some heavy lifting from highway vehicles—and they are up to the task. With cleaner cars and trucks, cleaner fuels, and better travel options, carbon emissions from cars and light-duty trucks can be cut by nearly 40 percent compared with 2005. Truck and bus emissions, which rose dramatically in the model scenario based on our current path, are held flat under the Blueprint despite the fact that the economy grows more than 80 percent.

drive up prices for fuels compared with the Reference case because Blueprint policies help drive down the price of oil.[90]

The one ironic impact of low gasoline prices is that they mute the ability of a cap-and-trade program to encourage consumers to purchase cleaner vehicles with better fuel economy, or to shift to other travel modes. Lower gas prices could therefore be seen as opening the door to more driving and urban sprawl. However, the Blueprint includes other policies that directly address those challenges, from limits on emissions from vehicles to per-mile driving fees, and those policies therefore deliver even more cost-effective cuts in carbon emissions (see Chapter 6).

7.6.6. Blueprint Case: Highway Vehicles Do the Heavy Lifting

The major Blueprint policies related to transportation focus on highway vehicles (cars, light-duty trucks, freight trucks, and buses). As a result, those vehicles deliver the majority of cuts in carbon emissions from transportation compared with the Reference case (see Figure 7.12).

Significant improvements in efficiency, cleaner fuels, and alternatives to today's travel patterns under the Blueprint allow cuts in carbon emissions from

cars and light-duty trucks of nearly 40 percent in 2030 versus 2005 (see Figure 7.15). That represents a significant improvement over the Reference case reduction of only 5 percent.

Trucks and buses face an even bigger task: under the Reference case their emissions rise nearly 40 percent. Under the Blueprint, their emissions remain flat despite the fact that the U.S. economy grows more than 80 percent.

7.6.7. Blueprint Case: Carbon Emissions from Air Travel Continue to Climb

Aircraft are the worst performer under the Blueprint, with their carbon emissions climbing more than 50 percent by 2030. The main Blueprint policy that affects the airline industry is the cap-and-trade system, as it puts a price on carbon emissions. Ironically, the overall success of the Blueprint policies keeps the resulting impact small: jet fuel prices are only 5–10 percent higher as a result of the cap, and do not really affect the use of air travel compared with the near doubling of jet fuel prices from 2005 to 2030 in the Reference case.

To reduce carbon emissions from air travel, our analysis includes only options for improving aircraft efficiency. Including other options could lead to greater reductions. For example, logistics changes such as

---

90   As with all savings on the cost of oil in this analysis, NEMS does not account for instability in the oil market, which could cause price spikes. The model also does not account for potential attempts by OPEC to reduce the oil supply and drive up prices in response to other nations' attempts to lower demand.

FIGURE 7.15. Changes in Transportation Carbon Emissions
(Blueprint case vs. Reference case)

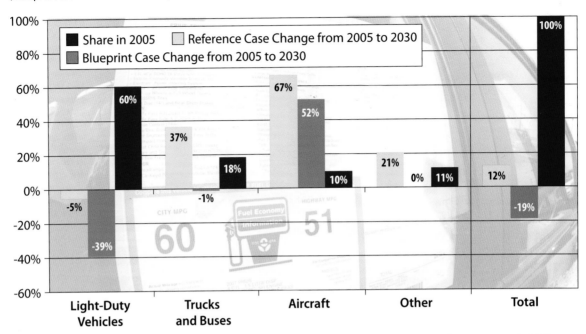

Transportation policies adopted under the Blueprint drive down carbon emissions nearly 20 percent in 2030 versus 2005. Cars and light-duty trucks lead the way with a near 40 percent reduction, while trucks and buses hold steady despite growth of 80 percent in the economy.

The carbon price under the Blueprint's cap-and-trade system encourages greater airplane efficiency but is too low to achieve bigger changes—these will require airplane carbon standards and cleaner fuels. Progress, however, can be made on the ground, where future airports can be more like the new Indianapolis airport terminal. Located in the middle of the airfield, planes taxi shorter distances and use less fuel. The terminal itself is also highly energy efficient, including, among other features, special windows that allow natural light and heat to enter the building in the winter and block it in the summer.

better routing to shorten flight distances, better scheduling to reduce congestion, and an update of the hub-and-spoke network (which relies on indirect stopovers and therefore increases fuel use) could have an impact.

High-speed electric rail can replace air travel between major commuting hubs, particularly in coastal regions and between major cities in America's heartland. However, large-scale investments in high-speed rail would have to accelerate significantly to affect global warming emissions by 2030. California will likely be the first state to build a high-speed electric rail system.

7.6.8. Blueprint Case: Low-Carbon Fuels Are on the Rise

Low-carbon biofuels and renewable electricity play important roles in our transportation Blueprint. Use of those fuels will rise to about 3.5 quadrillion Btu by 2030—or about 14 percent of all transportation fuel, and 20 percent of all highway fuel (see Figure 7.16). Much of this progress will occur because of the low-carbon biofuel portion of the renewable fuel standard included in our Reference case.

The low-carbon fuel standard in our Blueprint case takes that a step further by accelerating the phaseout of corn-based biofuels, which do not reduce carbon emissions and may even increase them during our time frame. The low-carbon fuel standard also drives a 10 percent increase in the efficiency of oil refineries, lowering carbon emissions from refineries by 1 percent per gallon of gasoline or diesel fuel made. The low-carbon fuel standard also ensures that high-carbon fuels such as liquid coal—which could double carbon emissions per gallon—do not make inroads and therefore undermine progress on curbing global warming.

While total electricity use in the transportation sector remains relatively modest under the Blueprint, it does grow rapidly from 2020 to 2030. In fact, the low-carbon fuel standard—along with the requirement that 20 percent of new light-duty vehicles be plug-ins or other electric-drive vehicles by 2030—drive a 10-fold increase in the use of electricity for transportation. Nearly 20 million plug-ins or other electric vehicles are on the road by 2030.[91]

And that progress is only the beginning for electric-drive vehicles. Under the Blueprint, the electricity sector does not tap the full potential for using renewable resources to generate power. That means that signifi-

FIGURE 7.16. Mix of Alternative Fuels under the Climate 2030 Blueprint

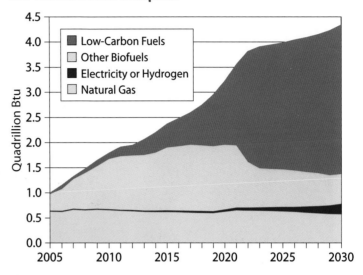

Alternatives to gasoline and diesel grow by more than a factor of four under the Blueprint. The renewable fuel standard in the 2007 Energy Independence and Security Act is responsible for much of this growth, but the Blueprint's low-carbon fuel standard helps shift the mix away from corn ethanol, which does not cut carbon emissions, and toward low-carbon alternatives such as electricity and cellulosic biofuels.

cant capacity is available to produce clean electricity for more plug-ins or battery-electric vehicles, or to produce hydrogen for use in fuel cell vehicles, as electric-drive vehicles dominate the car and light-truck markets beyond 2030.

7.6.9. Progress in Transportation Is Critical for Long-Term Success

While the transportation sector delivers significant cuts in carbon emissions under the Blueprint while saving the nation hundreds of billions of dollars, progress is still not as dramatic as in the electricity sector. Improvements in the latter will buy some time for progress in the transportation sector over the longer term.

However, that progress must begin today. The majority of benefits delivered under the Blueprint stem from solutions that have been available for a decade or more. Had the nation begun to phase in solutions such as more efficient vehicles, expanded transit, and reduced travel through per-mile pricing policies—and had we gotten serious about investing in low-carbon fuels and electric-drive vehicles two decades ago—many of the benefits of the Climate 2030 Blueprint would be available today.

---

91   The portfolio of potential advanced vehicles includes plug-in hybrids, battery-electric vehicles, and fuel cell vehicles. For ease of modeling, we used plug-ins as the sole technology, but other technologies with equal performance could be substituted.

BOX 7.5.

SUCCESS STORY

The Early Feats and Promising Future of Hybrid-Electric Vehicles

Hybrids combine the best elements of internal combustion engines with electric motors and batteries, and the best models can cut carbon emissions by 50 percent or more compared with conventional vehicles.

Hybrid-electric vehicles, which pair an internal combustion engine with one or more electric motors under the hood, first arrived in the United States 10 years ago. Since then the technology—and its popularity—have grown immensely.

Hybrids combine the best elements of internal combustion engines and electric motors to reduce carbon emissions by 30–50 percent compared with conventional vehicles, even while maintaining the performance, range, and other key features preferred by American drivers. While the hybrid concept is not new (patents were filed as far back as a century ago), it was only in the 1990s that batteries and onboard computers became advanced enough to permit successful hybrids.

The first modern, mass-produced hybrid was the Toyota Prius, a compact car brought to the Japanese market in 1997. Hybrids didn't reach the United States until three years later, when Honda unveiled its super-efficient two-seater Insight, followed promptly by Toyota with the Prius (Hall 2009). Both, however, were niche vehicles, with combined sales totaling less than one-tenth those of the top-selling model.

In 2004 hybrid technology finally reached a broader, mainstream U.S. audience. Toyota substantially redesigned its Prius, increasing not only its size but its fuel economy as well—an engineering feat that caught the attention of environmentalists and auto enthusiasts alike. Not to be outdone, Honda pushed its hybrid technology into the company's mainstream nameplates, releasing hybrid versions of the company's popular Civic and Accord sedans.

Annual U.S. sales steadily climbed as new models came to market, reaching 120,000 vehicles in 2005 (Ward's Auto Data n.d.). That same year Ford unveiled the Escape Hybrid as the first hybrid SUV. The range of consumer choices grew quickly: by 2006 the hybrid market consisted of 10 models representing five different vehicle classes. Today the U.S. hybrid market continues to expand. Sales climbed from roughly 20,000 in 2001 to more than 300,000 in 2008, with the Prius now ranked among the top-10 best-selling vehicles in the country (Ward's Auto Data n.d.).

That said, not all hybrid models have been successful. Honda abandoned its Accord Hybrid in 2006 (AP 2007);

sales for Toyota's Lexus brand "performance hybrids" have flagged; and Chrysler discontinued its Durango and Aspen hybrids after their first year (Doggett and O'Dell 2008). The critical difference between the hybrid standouts and the hybrid also-rans is this: hybrid vehicles that use the technology to boost power rather than increase fuel economy have failed to capture significant market share.

Responding to consumer preference, automakers are now moving their hybrid vehicles toward efficiency. Honda's 2009 Insight (a new, larger sedan bearing very little resemblance to its discontinued two-seater namesake), for example, will be a 40-plus-mpg vehicle selling for less than $20,000 (Honda 2009). Ford is entering the hybrid car market with a Fusion Hybrid in 2009 that offers better fuel economy than its midsize competitor, the Toyota Camry Hybrid. And Toyota is bringing out its third-generation Prius with an expected 50-mile-per-gallon fuel economy rating (Kiley 2008).

The next few years will likely see an even greater revolution in hybrid design, with major-manufacturer release of plug-in hybrid-electric vehicles (PHEVs). Plug-ins, as they're commonly known, have battery packs large enough to enable drivers to travel significant distances on electric power alone, and to recharge the vehicles at home through conventional power outlets. Yet their use of gasoline engines also allows the vehicles to meet consumers' requirements for range and refueling time.

In short, PHEV designs provide an overall improvement in fuel economy and the opportunity—with a clean-power grid—to dramatically reduce vehicles' carbon emissions. General Motors, Toyota, and Ford are slated to bring the first mass-produced plug-ins to market between 2010 and 2012. Although cost and battery-engineering challenges remain, a cleaner vehicle future looks promising.

The year 2030 should be viewed as a critical milepost on the path to reducing global warming emissions 80 percent or more by 2050. If transportation policies do not provide the cuts we outline by 2030, the nation has little chance of reaching the 2050 target.

7.7. Land-Use Implications of the Blueprint

Some Blueprint solutions, such as an increase in renewable electricity, the use of biofuels, and carbon offsets from agriculture and forests, have implications for land use. At the same time, a move away from heavy reliance on fossil-fuel-based energy will provide significant land-use benefits. This section outlines some of the key land-use implications of the Blueprint solutions.

We recognize that the use of land to reduce global warming emissions may inadvertently create new environmental or sustainability problems, economic effects such as higher prices for agricultural commodities, and even an indirect increase in heat-trapping emissions. We have deliberately restricted the kind and level of certain solutions, such as bioenergy and offsets, to minimize the possibility that the nation will divert land from productive uses and indirectly create adverse effects on land use in other countries.

7.7.1. Land Use and Energy under the Blueprint

While expanding the use of renewable electricity and biofuels will have important effects on land use, it will also reduce the effects on land use of producing, transporting, and using fossil fuels. The environmental impacts of using land to produce and burn fossil fuels tend to be much greater than those of producing renewable energy and storing carbon in soils and trees.

Under the Blueprint, the total land area needed to produce electricity from wind and solar power is 1,500–36,600 square miles, or about 0.04–1 percent of all U.S. land area. The low end of the range includes only the footprint of wind turbines and their supporting infrastructure and large-scale solar projects. It does not include the area occupied by distributed PV, which is typically installed on residential and commercial buildings, and therefore would not require any new land. The high end of the range includes both the footprint of the turbines and the land between them, which could still be used to grow crops or graze animals, as well as the area used by distributed PV.

By 2030, the cumulative reduction in coal use from increased energy efficiency and renewable energy under the Blueprint would result in nearly 13,000 square miles of avoided land use from both surface and mountaintop-removal coal mining. While state and

Under the Blueprint, the land needed to produce clean electricity from wind and solar power and transportation fuels from corn and switchgrass—as well as the land needed to sequester carbon in new forests and smart agricultural practices—would only require 2 to 6 percent of the total U.S. land area. At the same time, the Blueprint would significantly reduce the amount of land used for coal mining, oil and natural gas drilling, fossil fuel power plants, refineries, pipelines, waste disposal, and related infrastructure.

(see Figure 7.17). Energy crops would require about 39,000 square miles (25 million acres) of land by 2030, or about 2.5 percent of all land now used for agriculture in the United States. Most energy crops would be grown on pasture lands and would be used to produce cellulosic biofuels.

Overall, a reduction in land used for oil and natural gas drilling, fossil fuel power plants, refineries, pipelines, waste disposal, and related infrastructure would offset some of the increase in the amount of land used for renewable electricity and biofuels under the Blueprint. However, our analysis did not quantify these land-use benefits.

7.7.2. Land for Carbon Offsets from Agriculture and Forestry

All the domestic carbon offsets we modeled come from a category in the NEMS model called "biogenic sequestration offsets." These offsets are generated through increased storage of carbon in soils and vegetation in the agriculture and forestry sectors—primarily as a result of changes in soil management practices, better forest management, and afforestation. Of those, afforestation is the only one that would require diverting new land for this specific purpose. The other strategies involve changing practices on lands already used for the same purpose.

Although NEMS does not show what percentage of offsets stem from afforestation, we can try to estimate that percentage based on information from the Environmental Protection Agency's Forest and Agriculture Sector Optimization Model with Greenhouse Gases (FASOMGHG) (Murray et al. 2005).

Extrapolating from data from existing runs of FASOMGHG, at the carbon prices shown in the Blueprint, we estimate that roughly 50 percent of the domestic offsets in our results are likely to come from afforestation.[92] That means that of the 314 million metric tons of domestic offsets used by capped firms in 2030, 157 million metric tons come from afforestation.

Based on estimates of the amount of carbon that afforestation stores per acre (Birdsey 1996), we estimate that the added land area needed to sequester the 157 million metric tons in 2030 would range from 17 million to 71 million acres (with a midpoint of 44 million acres). Most of this afforestation would likely occur on marginal croplands, grasslands, and rangelands. The midpoint estimate of 44 million acres represents

federal laws require reclamation of land permitted for coal mining, in practice the coal industry has reclaimed only a small portion of this land. And in many cases—particularly for mountaintop-removal mining—the reclaimed land does not resemble its original state.

Growing energy crops (switchgrass) and corn to produce biofuels for transportation under the Blueprint would require more than 52,000 square miles of land in 2030—or about 1.4 percent of all U.S. land area

---

92 This is a rough approximation based on extrapolation from existing model results in Murray et al. 2005. We did not conduct any new runs of the FASOMGHG model.

BOX 7.6.

Impact of the Blueprint Policies in 2020

A central insight from the Blueprint analysis is that the nation has many opportunities for making cost-effective cuts in carbon emissions in the next 10 years (through 2020). Our analysis shows that firms subject to the cap on emissions find it cost-effective to cut emissions more than required—and to bank carbon allowances for future years. Energy efficiency, renewable energy, reduced vehicle travel, and carbon offsets all contribute to these significant near-term reductions.

By 2020, we find that the United States can:

- Achieve, and go beyond, the cap requirement of a 26 percent reduction in emissions below 2005 levels, at a net annual savings of $240 billion to consumers and businesses. The reductions in excess of the cap are banked by firms for their use in later years to comply with the cap and lower costs.

- Reduce annual energy use by 17 percent compared with the Reference case levels.

- Cut the use of oil and other petroleum products by 3.4 million barrels per day compared with 2005, reducing imports to 50 percent of our needs.

- Reduce annual electricity generation by almost 20 percent compared with the Reference case while producing 10 percent of the remaining electricity with combined heat and power and 20 percent with renewable energy resources, such as wind, solar, geothermal, and bioenergy.

- Rely on complementary policies to deliver cost-effective energy efficiency, conservation, and renewable energy solutions. Excluding those energy and transportation sector policies from the Blueprint would reduce net cumulative consumer savings through 2020 from $781 billion to $602 billion.

---

about 4 percent of all cropland, grassland pasture, and rangeland in the United States.

7.8. Sectoral Policies Are Essential for a Cost-Effective Blueprint

The Blueprint analysis reveals the benefits of pursuing complementary policies along with a cap-and-trade program. A cap on carbon emissions is critical because it establishes the level of cuts in global warming emissions regardless of the rest of the policy mix. However, adding sector-based policies helps deliver those reductions in a more cost-effective way. We demonstrated this finding by developing a sensitivity, or No Complementary Policies, case: that is, by running the model while excluding all the sector-based policies from the Blueprint.

As noted, because of limitations in the NEMS model, we were unable to model a critical feature that would help make a cap-and-trade program more cost-effective: namely, we could not target revenues from the auction of carbon allowances for specific purposes such

as funding energy efficiency measures and low-carbon technologies. Instead, we could only assume that government recycles such revenues back into the economy, to consumers and businesses.

With that limitation in mind, our results show that if we exclude all sector-based policies, we are left with only a price for carbon emissions to drive global warming solutions into the marketplace. A carbon price alone will change the energy and technology mix and spur some improvements in energy efficiency and conservation. However, it will not provide all the needed cost-effective solutions because of other market barriers, such as consumers' aversion to risk and the up-front cost of more advanced technology (see Chapters 4–6). Sector-based policies are critical to overcoming those barriers, facilitating the development and deployment of clean and efficient technologies, and delivering them at a lower cost than a carbon price alone could do.

The next sections explore some of the findings of the No Complementary Policies case.

FIGURE 7.17. Total Land-Use Effects of Renewable Electricity, Biofuels, Avoided Coal Mining, and Afforestation Offsets in 2030

(Blueprint case)

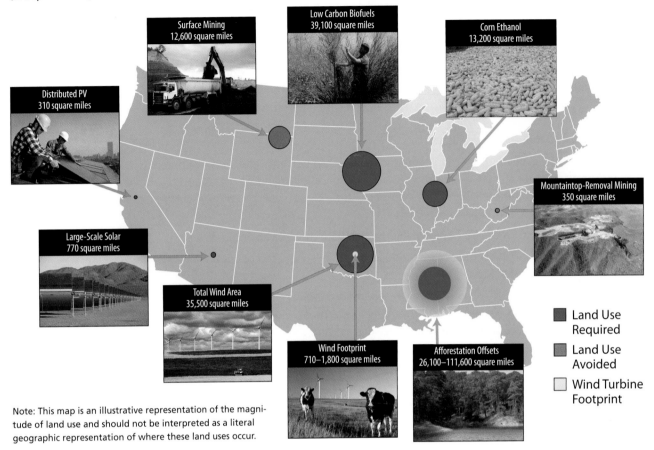

Note: This map is an illustrative representation of the magnitude of land use and should not be interpreted as a literal geographic representation of where these land uses occur.

7.8.1. No Complementary Policies Case: Impact on Prices of Carbon Allowances

A comparison of the results of the Blueprint case with those of the sensitivity case shows that stripping out the complementary policies leaves a basic cap-and-trade system without targeted recycling of revenues—and that the prices of carbon allowances more than double (see Figure 7.18). The lower prices of allowances under the complete Blueprint allow consumers to see much smaller increases in the rates they pay for electricity and fuels.

Each sector's policies play a significant role in cutting the prices of carbon allowances. With the transportation sector's policies stripped out, allowance prices rise by about 33 percent. If we also strip out policies related to the electricity, industry, and buildings sectors, allowance prices rise by almost another 66 percent.[93]

The reason for these lower prices is straightforward: energy efficiency, clean technologies, and conservation play far more significant roles in our Blueprint results than would be possible with only a carbon price signal, and encourage the adoption of cost-effective solutions that have a dampening effect on the prices of both allowances and fuel. Energy efficiency technologies cost more up-front, so risk-averse consumers can be more reluctant to purchase them despite the long-term financial savings they can provide.[94]

(continued on page 160)

93 In our sensitivity case, we stripped out the transportation policies first and then stripped out the other policies. Had we stripped out the policies related to electricity, industry, and buildings first and those related to transportation second, the changes in allowance prices might have been different.

94 The Congressional Budget Office estimates that fuel prices would need to rise by 46 cents per gallon to reduce gasoline use by 10 percent (CBO 2004).

TABLE 7.5. Land Needed for Renewable Electricity, Biofuels, and Afforestation Offsets (2030)

Technology	Land Area:[a] Increase over Reference Case (square miles)	Land Area:[a] Total New plus Existing (square miles)	Percent of Total U.S. Land Area
Electricity			
Total Area for Wind[b]	16,341	35,466	1.0%
Wind Footprint[b]	327–817	709–1,773	0.02–0.05%
Central Photovoltaics	122	126	0.004%
Distributed Photovoltaics[c]	78	312	0.01%
Concentrating Solar Thermal	482	647	0.02%
Electricity Subtotal[d]	931–17,023	1,482–36,551	0.04%–1.03%
Low-Carbon Biofuels[e]	0	39,063	1.10%
Corn Ethanol[f]	– 33	13,160	0.37%
Afforestation Offsets[g]	26,121–111,608	26,121–111,608	0.74–3.16%
Total	**27,019–128,598**	**79,826–200,382**	**2.26–5.66%**

Notes:

a The incremental land area is based on the increase in renewable electricity, biofuels, and afforestation under the Blueprint compared with the Reference case. The total land area is based on both existing and new renewable electricity, biofuels, and afforestation in 2030 under the Blueprint. See Appendix D online for assumptions and sources.

b The wind footprint includes the land used by the wind tower base, access roads, and supporting infrastructure. The total for wind includes the footprint as well as the area between the turbines that can be used for other productive uses, such as farming.

c Distributed photovoltaics are installed on residential and commercial buildings, and therefore would not require any new land.

d The low end of the range includes only the wind footprint and does not include distributed PV, while the high end of the range includes the total areas for wind and distributed PV.

e These figures are based on an estimate of the amount of energy crops (switchgrass) used for producing biofuels. The incremental land area is zero under the Blueprint because no additional cellulosic biofuels are produced above the Reference case. We assumed that the use of agricultural, forest, urban, and mill residues would not require any new land, as these residues come from existing operations.

f Land use for corn ethanol reaches a maximum of 31 million acres, or 40 percent of the total corn crop, in 2017, and then declines to 8.4 million acres, or 11 percent of the total corn crop, by 2030, as lower-cost cellulosic biofuels replace corn ethanol.

g The land for afforestation offsets is based on the assumption that 50 percent of the total offsets in the Blueprint cap-and-trade program come from afforestation, and assumes carbon storage of 2.2 million to 9.4 million tons of CO_2eq per acre per year for up to 120 years.

FIGURE 7.18. Prices of Carbon Allowances

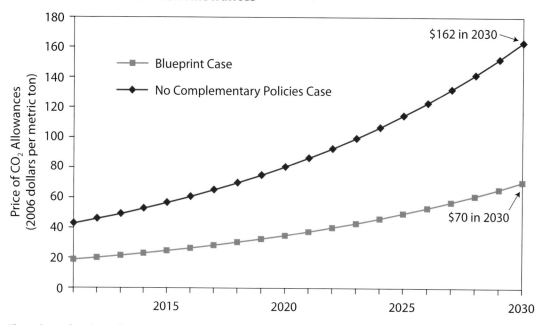

The prices of carbon allowances are at least twice as high in the No Complementary Policies case as in the Blueprint case.

Blueprint policies that promote energy efficiency and cleaner technologies will significantly increase the prevalence of "green" buildings in the United States. The LEED Gold-certified City Hall in Austin, TX, employs daylighting, occupancy-controlled lighting sensors, and efficient appliances to reduce electricity demand; a high-efficiency natural gas boiler for hot water and heating; and a district cooling system that saves energy during peak daytime hours.

FIGURE 7.19. The Source of Cuts in Global Warming Emissions in 2030

(No Complementary Policies case vs. Reference case)

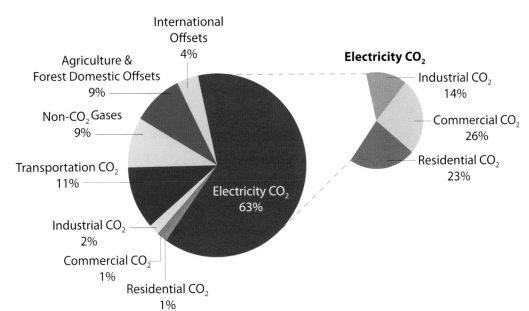

The electricity sector leads the way in cutting emissions, playing an even larger role than under the Blueprint case. Offsets follow, and also play a larger role than in the Blueprint case. Transportation is third, playing a smaller role than under the Blueprint. Emission cuts in the electricity sector include reductions in demand from energy efficiency in the residential, commercial, and industrial sectors.

FIGURE 7.20. U.S. Energy Use

(Blueprint case vs. No Complementary Policies case)

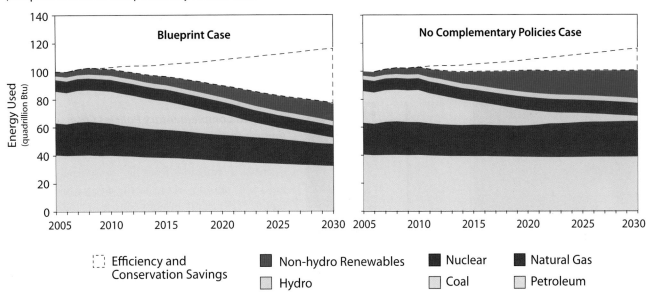

In the No Complementary Policies case, energy efficiency and conservation play a much smaller role in reducing U.S. energy use, while renewable energy, natural gas, carbon capture and storage, and nuclear power play a larger role in the electricity sector. Oil use is also greater without the cleaner cars and trucks and better transportation choices delivered by the Blueprint's complementary policies.

FIGURE 7.21. Net Cumulative Savings (2010–2030)

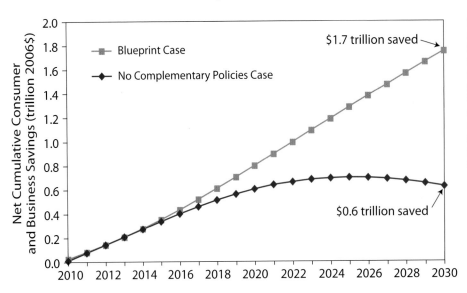

The 2010–2030 net cumulative savings to consumers and businesses are $1.7 trillion under the Blueprint case. Under the No Complementary Policies case, which strips out all the energy and transportation policies, these savings are $0.6 trillion.

Note: Net present value using a 7% real discount rate.

The complementary policies significantly increase the use of energy-efficient technologies in the building, commercial, and industrial sectors. These policies also expand the use of cleaner cars and trucks, and lower demand for travel, more than the carbon price signal alone.

The complementary policies have the added benefit of moving important technologies into the marketplace early, advancing them up the learning curve, bringing down their costs, and continuing to provide benefits beyond 2030. Funding for research and development will also help bring new breakthrough technologies to the market more quickly. Wide-scale deployment of all these low-carbon technologies cannot happen overnight, so any significant delays could eliminate the nation's chances of cutting carbon emissions 80 percent by 2050.

7.8.2. No Complementary Policies Case: Impact on Cuts in Carbon Emissions

Excluding the complementary policies also shifts reductions in emissions from the transportation, buildings, and industry sectors to the electricity sector (see Figure 7.19). Cuts in emissions from the electricity sector in 2030 grow from 57 percent under the Blueprint case to 63 percent in the No Complementary Policies case.

Carbon offsets play a slightly larger role in 2030, accounting for 13 percent of total cuts in emissions versus 11 percent under the Blueprint. On the other hand, cuts in emissions from the transportation sector drop to 11 percent of the total, versus 16 percent under the Blueprint. Reductions in non-CO_2 gases contribute 9 percent of the cuts.

7.8.3. No Complementary Policies Case: Impact on Total Energy Use

Our sensitivity results show that, without the help of complementary policies, the most important lowest-cost

Many of the complementary policies in the Climate 2030 Blueprint aim to help increase the use of energy-efficient technologies in our homes, offices, and factories. Energy efficiency is one of the most cost-effective sources of emissions reductions in the Blueprint.

solutions—energy efficiency and conservation—play a much smaller role (see Figure 7.20). That is because a carbon price signal alone cannot overcome significant market barriers to investments in energy efficiency and conservation.

Strong climate policies that help the nation transition to a more efficient, cleaner, low-carbon economy will help us avert some of the worst consequences of global warming.

The result is that renewable energy, natural gas, carbon capture and storage, and nuclear power play a larger role in the electricity sector. The renewable energy mix also includes more higher-cost choices such as

offshore wind, dedicated biomass plants, advanced geothermal, and solar, which all become more competitive than other low-carbon options at higher prices for carbon allowances. Oil use is also greater without cleaner cars and trucks and reduced travel in the transportation sector.

7.8.4. No Complementary Policies Case: Impact on Consumer Savings

With the complementary policies stripped out, cumulative net consumer and business savings are lower than in the Blueprint case. In the sensitivity analysis, cumulative consumer and business savings reach $0.6 trillion in 2030, compared with $1.7 trillion with the complementary policies in place (in 2006 dollars with a 7 percent discount rate) (see Figure 7.21). These comparisons assume that government recycles revenues from the auction of carbon allowances back into the economy, but does not target those revenues to specific uses such as energy efficiency and low-carbon technologies.

Our results show that putting a price tag on carbon emissions is insufficient to overcome market barriers that hinder the growth of the most important and least expensive climate solution: energy efficiency. Targeted policies encourage up-front investments in energy-saving technologies such as those found at the West Grove, PA, headquarters of Dansko Inc., a shoe manufacturer. Rooftop storm water collection (for use in the building's toilets), insulated windows, solar hot water heating, a green roof, and solar shades earned Dansko LEED Gold certification.

Climate 2030 Blueprint policies will benefit the U.S. economy by spurring investments in clean energy technologies and saving consumers and businesses money—in every region of the country.

The Climate 2030 Blueprint will jump-start a clean energy transition that will stabilize energy prices and put money in consumers' pocketbooks.

7.9. Economic, Energy, Health, and Global Benefits of Strong U.S. Climate Policies

Strong climate policies that help the nation transition to a more efficient, cleaner, low-carbon economy will not only help us avert some of the worst consequences of global warming. Such policies will also provide a host of other benefits, including opportunities for economic growth, more stable sources of energy, reductions in other pollutants, improvements in public health, and opportunities for cooperation and development worldwide.

Economic benefits. Climate policies will give a boost to our economy by providing new jobs in the clean technology sector, spurring technological innovation, creating opportunities to export those technologies, and stabilizing energy prices. Several recent studies show that this "green transition" will create millions of well-paying jobs (Apollo Alliance 2008; Pollin et al. 2008).

A recent UCS analysis of a 25 percent national renewable electricity standard by 2025 showed that this policy alone would create 297,000 new jobs in 2025—or more than three times as many jobs created by producing an equivalent amount of electricity from fossil fuels (UCS 2009). Renewable energy creates more jobs than fossil-fuel-based energy because it is typically more labor-intensive.

Energy benefits. Volatile energy prices and uncertainty about future sources of energy play havoc with our economic well-being. By taking advantage of the huge potential of energy efficiency, and by transitioning our energy supply to clean, reliable, renewable sources, we can help stabilize energy prices and improve the long-term health of our economy.

Reductions in other pollutants and improvements in public health. Production and consumption of goods and services often result in other forms of pollution besides carbon emissions. For example, burning fossil fuels releases sulfur emissions, mercury emissions, and particulate matter, among other harmful co-pollutants. Mining, drilling, transportation, and waste disposal related to coal, oil, natural gas, and nuclear power also pose serious health and environmental hazards. By implementing policies that cut our carbon emissions, we can also reduce these other pollutants (see Box 7.3).

Global cooperation. Climate change is a global problem, and all nations will need to take serious action to address it. However, the United States has a unique responsibility to play a leadership role in curbing global warming because of the outsized volume of our past and current heat-trapping emissions, and the wealth we built on those emissions.[95]

---

95 At the 1992 Earth Summit in Rio de Janeiro, 177 countries including the United States signed the U.N. Framework Convention on Climate Change. That framework clearly recognizes "common but differentiated responsibilities" among the signatory nations, and assigns the lead responsibility to developed countries.

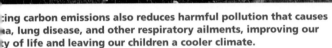

:ing carbon emissions also reduces harmful pollution that causes
a, lung disease, and other respiratory ailments, improving our
ty of life and leaving our children a cooler climate.

Aggressive U.S. action to reduce carbon emissions will send a strong
signal to the rest of the world that we must work together to tackle
this global problem.

The most important step we can take is to make a strong commitment to reducing our carbon emissions. As our analysis shows, our nation will reap tremendous benefits from doing so. We cannot solve global warming on our own, but our leadership will set the stage for other countries to take critical steps to reduce their emissions as well.

7.10. Limitations, Uncertainties, and Opportunities for Future Research

Projections of long-term changes in the supply, use, and prices of energy are subject to a great deal of uncertainty. Modeling the impacts of climate and energy policies that will require significant changes in the way we produce and use energy only adds to those uncertainties.

One limitation of our analysis is that we analyzed only two potential scenarios for meeting our targets for reducing global warming emissions. Other scenarios with different policy, economic, and technology assumptions could achieve these or more stringent targets, with different effects.

The most important types of assumptions we made concerned:

- energy demand and prices;
- the cost and performance of technologies;
- trajectories for emissions set by the cap, levels of offsets, and a zero terminal balance in the allowance bank;

- levels of development and policies for energy efficiency, conservation, and renewable energy;
- the availability and cost of carbon capture and storage, advanced nuclear power plants, and emerging renewable energy and transportation technologies; and
- the amounts of biomass available to provide electricity and fuels.

We were also unable to address a variety of limitations of NEMS, despite incorporating information from other analyses and modifying the model. Examples are described below.

Limitations of macroeconomic modeling. NEMS has significant limitations in how it quantifies the macroeconomic impacts of climate and energy policies. For example, it cannot fully account for the positive effects on GDP and employment of investments in energy efficiency, renewable energy, and other low-carbon technologies, and of savings on consumers' energy bills.

Indeed, NEMS predicts roughly the same gain in economic productivity in the Reference case as in the Blueprint case. That result understates the nation's ability to shift savings from reduced energy use to more productive uses. Nor does NEMS value other productivity gains and non-energy benefits that would both accelerate adoption of more advanced technologies and improve economic performance (Worrell et al. 2003).

Recent research by both utility companies and government agencies suggests that wind power can contribute up to 25 percent of the U.S. electricity supply without requiring storage or compromising the reliability of the electricity grid.

The model also treats reductions in energy consumption and increases in energy prices as exerting a negative impact on the economy, even if overall energy bills are lower. And NEMS does not account for the loss of GDP that may result from unchecked climate change in the Reference case.

As noted, the model is also not designed to target allowance revenues to specific technologies and purposes in ways that could reduce carbon emissions and improve economic welfare. Although the model can recycle these revenues generally to households, businesses, and government, modifying the model to include a more extensive approach was beyond the scope of our study.

Modeling energy efficiency. The model does include specific technologies for boosting energy efficiency in vehicles, industry, and buildings. However, analyzing the impact of proposed efficiency policies in the residential, commercial, and industry sectors is difficult without significantly modifying the model and its assumptions.

The model does attempt to capture some reductions in energy use owing to higher prices. However, that approach is limited. The way NEMS shows consumers and businesses adopting technologies in response to

changes in price depends on fixed elasticities, or payback times linked to specific discount rates. But those elasticities and payback periods can shift over time because of changes in household income or consumer preferences.

As preferences evolve and as consumers become more aware of choices, the resulting carbon price signal needed to drive those choices may be substantially lower than NEMS might indicate. Stated differently, changed behaviors may deliver greater efficiencies or reductions in emissions for the same price signal.

The effects of sources of electricity with variable power output. The model does not fully capture the impact of high levels of variable-output wind and solar on the electricity grid. NEMS does capture variations in the output of these technologies during nine different time periods throughout the year for 13 different U.S. regions. However, it does not capture all the fluctuations that can occur over much shorter time periods, and at the subregional level. Doing so would require additional ramping up and down of other sources of power.

Several studies by U.S. and European utilities and government agencies have found that wind can capture as much as 25 percent of the electricity market at a modest cost, and without adverse effects on the system's reliability or the need to store power (EERE 2008; Holttinen et al. 2007). Our results are below these levels, with wind and PV capturing about 20 percent of the U.S. electricity market by 2030.

Offshore carbon emissions. The NEMS model does not track changes in heat-trapping emissions in other countries from the production of energy and other goods imported into the United States.

This shortcoming is significant for the transportation sector, which is responsible for the majority of the 3.5-million-barrel-per-day cut in imported petroleum products in the Blueprint versus the Reference case. Given projections that the United States could import more than 6 million barrels a day of high-carbon resources such as tar sands and oil shale by 2035 (Task Force 2006), our results could overlook significant cuts in carbon emissions that could result from curbing reliance on those overseas resources.

For example, if the 3.5 million barrels per day came from tar sands, U.S. cuts in global warming emissions in 2030 would rise by 2 percent under the Blueprint.[96] If the 3.5 million barrels per day came from oil shale, projected cuts in emissions could rise by 15 percent.[97]

NEMS also does not include carbon emissions from indirect changes in land use, either domestically or abroad, that could occur from using food crops and certain agricultural land to produce biofuels. Some estimates show that such indirect effects from the use of corn as a biofuel feedstock could nearly double carbon emissions compared with gasoline (Searchinger et al. 2008). Our Blueprint findings may therefore underestimate the benefit of moving away from corn ethanol. We have tried to minimize displacement of U.S. agricultural crops (see Section 7.7) to prevent potential adverse effects abroad, such as the clearing of rainforests to produce crops formerly grown in the United States.

These limitations of our model, and the uncertainties around some of our key assumptions, present important opportunities for future research. Different combinations of technologies and policies could also be modeled. Another important extension could be to more fully examine the effects of Blueprint policies on employment and other aspects of the economy.

---

96   This assumes that global warming emissions from fuel from tar sands would be about 15 percent higher than those from today's gasoline, on a well-to-wheels basis.

97   This assumes that global warming emissions from fuel from oil shale would be about double those from today's gasoline, on a well-to-wheels basis.

Cultivating a Cooler Climate:
Solutions That Tap Our Forests and Farmland

How we manage U.S. forests and farmlands has a major impact on our net emissions of carbon dioxide and other heat-trapping gases. The United States has a rich diversity of forests, from the maple-beech-birch woodlands of New England to the lob-lolly pinelands of the Southeast to the coastal redwoods of northern California. Covering almost 750 million acres of public and private lands, our forests provide critical habitat for wildlife, as well as recreational opportunities, sources of fresh water and timber, and aesthetic benefits for millions of people.

These forests are also important storehouses of carbon, with some 245 million metric tons carbon dioxide equivalent (MMTCO$_2$eq) stored in living tissue, leaf litter, and forest soils (CCSP 2007). Through photosynthesis, trees and other vegetation take up—or sequester—carbon. A combination of natural disturbances and human activities, including timber harvests, fire, pest infestations, and deforestation also release carbon back into the atmosphere as carbon dioxide. Today U.S. forests are a net "sink" for carbon, drawing more CO$_2$ out of the atmosphere than they release.

The United States is also home to some 1,400 million acres of cropland and grazing lands. Agriculture is a complex, malleable enterprise with variable impacts on net global warming emissions. Major sources of CO$_2$ include soil disruption, such as through tillage for crops; the fossil-fuel-intensive production of herbicides, insecticides, and, especially, industrial fertilizers; and the use of fuel to run farm machinery. Besides CO$_2$, agricultural activities allow the release of two other potent heat-trapping gasses, methane and nitrous oxide, from livestock, manure, and nitrogen fertilizers applied to soils.

Land-management practices and policies exert a major impact on U.S. heat-trapping emissions. In 2000 the United States emitted more than 7,000 MMTCO$_2$eq. The great majority of those emissions stemmed from the burning of fossil fuels, including 50 MMTCO$_2$eq from on-farm use. U.S. forests, in contrast, were a major net sink of carbon, absorbing almost 840 MMTCO$_2$eq, or about 12 percent of U.S. emissions in 2000.

We are heading in the wrong direction. Our forests and other vegetation absorb more than 10 percent of U.S. global warming emissions, but that capacity is at substantial risk. More than 50 million acres of undeveloped, privately held lands are projected to be converted to urban and developed uses over the next 50 years (USFS 2007).

A recent EPA study projected that, under business as usual, the U.S. forest carbon sink will decline to about 220 MMTCO$_2$eq by 2020, and 145 MMTCO$_2$eq by 2030, with emissions from farmlands projected to remain high (Murray et al. 2005). Together forests and farms in the continental United States will soon become a major net source of emissions, contributing a projected 280 MMTCO$_2$eq to the atmosphere in 2020, and 320 MMTCO$_2$eq in 2030, from non-fossil-fuel sources. Without a major course correction, our lands will amplify—rather than reduce—global warming emissions.

We can do better. The Congressional Budget Office estimates that U.S. forests and farmlands have the technical (biophysical) potential to sequester the equivalent of 13–20 percent of what the nation's CO$_2$ emissions were in 2005, through expansion of forests onto lands now under other uses, reduced deforestation, and better management of current forests and farmlands (CBO 2007). Barriers to realizing that potential include the costs of altering land-use practices, trade-offs between carbon mitigation and other social goals, and the potential for climate change to reduce carbon storage by increasing the frequency and severity of fire and pest infestations in some U.S. forests (van Mantgem et al. 2009).

Recent modeling studies show that privately held U.S. forests and farmlands have the potential to cost-effectively sequester substantial quantities of CO$_2$ over the next few decades (CBO 2007; Murray et al. 2005), particularly through accelerated planting of trees on non-forest lands in the Midwest and Southeast. Further research is needed to refine these projections, to account for competing land uses, and to ensure that any expansion of forests (and bio-fuels) onto lands now used for food crops does not

Although not modeled in the Climate 2030 Blueprint, our forests have an important role to play in reducing global warming. Carbon is captured by trees during photosynthesis and stored in living tissue, leaf litter, and forest soils. Government projections, however, suggest that our forests' capacity to absorb global warming emissions could decline rapidly in just two decades. This dangerous trend can be addressed by encouraging better forest management on public lands, providing tax incentives for owners of private forests, and implementing land-use plans that enhance the capacity of lands to store carbon.

increase the price of agricultural commodities or raise emissions because of land-use changes in other countries (Searchinger et al. 2008). Research is also needed to develop robust estimates of the amount of carbon that the more than 40 percent of U.S. forestlands that are publicly owned could absorb (Smith and Heath 2004).

Smart Policies and Practices

Global warming solutions for U.S. forests. Most policy debates on how to boost carbon storage in forests have focused on carbon offsets under a cap-and-trade program. However, because large-scale reliance on offsets could enable capped companies to avoid cutting their emissions, leaders at the federal, state, and local levels need to develop a broader portfolio of policies designed to inventory and expand the amount of carbon forests store, and enhance the other critical benefits they provide.

The federal government, for example, could more fully integrate carbon storage into the man-

agement goals for 182 million acres of federally owned forests in the continental United States. The government could also require longer rotations for timber harvests, the use of reduced-impact harvesting techniques, and better management of fires and pests on public lands.

Federal and state governments can also provide tax incentives to owners of private forests who increase carbon storage on their lands, and offer challenge grants to communities to plant trees and pursue other programs that conserve carbon. Land-use plans, zoning ordinances, and laws protecting natural resources are all tools that local governments can use to encourage smart growth, protect open space, and maintain and enhance the capacity of lands to conserve carbon (Stein et al. 2008).

Global warming solutions for U.S. farmlands. Agriculture is such a complex and varied enterprise that its implications for global warming, like those of forestry, are best addressed through an integrated

More than half of the carbon dioxide humans put into the atmosphere each year is soaked up by trees, crops, soils, and oceans, helping to slow global warming. In addition to carbon storehouses, our forests are sources of fresh water and timber, homes for wildlife, and places of beauty and recreation.

products, and efforts to inform farmers about better feed mixes. The government could also investigate alternatives to anaerobic systems for storing manure, like hog lagoons, that produce methane. And it could provide incentives for the use of methane digesters to capture methane as a source of on-farm energy. Programs for promoting biofuels could emphasize the planting of deep-rooted grasses, which can enhance carbon sequestration as well as offset the use of fossil fuels.

Central to a climate-friendly policy agenda should be the promotion of agricultural systems that provide multiple climate as well as other environmental benefits. Organic-style cropping systems, for example, avoid the use of fossil-fuel-intensive insecticides, herbicides, and chemical fertilizers by employing multiyear rotations to suppress pests, and conservation tillage to suppress weeds (see Box 7.7). These systems also avoid the need for industrial nitrogen by relying on nitrogen-fixing cover crops to keep the soil fertile and promote the buildup of organic matter, maximizing the soil's potential as a carbon sink. The government can encourage the use of these systems through programs that provide transition and cost-share assistance to farmers.

This new policy agenda should rest on long-term, multidisciplinary research illuminating the connections between agricultural practices and heat-trapping emissions. Recent studies, for example, cast doubt on the efficacy of no-till—a kind of conservation tillage—in sequestering carbon (Blanco-Canqui and Lal 2008; Baker et al. 2007). Well-designed research will help reveal the inevitable trade-offs and new opportunities implicit in our choice of agricultural practices.

Important research also includes more detailed studies of the carbon-storing effects of conservation and no-till agriculture, differences between grain-based confinement livestock/poultry systems and pasture-based systems, better ways to replace chemical fertilizers with animal waste, and the factors that lead to the release of nitrous oxide from agricultural systems.

Through these and other smart policies and practices, the nation can fully realize the potential of our forests and farmlands to cultivate a cooler climate while providing other goods and services on which we depend.

set of policies and programs, some new and some already in place. The federal government, for example, could expand the Conservation Reserve Program to encourage farmers to sequester carbon, and maximize incentives under the Conservation Security Program for the use of cover crops, crop rotation, conservation tillage, and other carbon-conserving practices.

New programs to reduce methane emissions from cattle could include educational campaigns to discourage the consumption of beef and dairy

BOX 7.7.

SUCCESS STORY

Farmers and Fungi: Climate Change Heroes at the Rodale Institute

Set amid the gently rolling hills of southeastern Pennsylvania, a little miracle unfolds every day. The miracle workers are microscopic fungi that live inside and around the roots of crops, holding the fabric of fertile soil particles together and simultaneously storing carbon.

Located on a 333-acre certified organic farm in Kutztown, PA, the Rodale Institute has been studying organic farming methods for more than six decades (Rodale 2009). Of particular interest, the institute has overseen a side-by-side comparison of organic and conventional farming practices since 1981. The longest-running experiment of its kind, this study shows that organic agricultural practices are regenerative—that is, they build the soil. Farmers practicing regenerative organic agriculture plant cover crops; rotate crops; avoid herbicides, insecticides, and industrial fertilizers; and fertilize with composted manure. Done in a smart way, these practices rebuild poor soils, use nitrogen efficiently, and remove carbon from the air and store it in the soil, where it accumulates year after year.

in turn captures carbon dioxide from the atmosphere and stores it underground.

Enter those amazing fungi. Allowed to flourish, mycorrhizal fungi perform two important functions: they help slow the decay of organic matter, and they

This roller-crimper tractor attachment is designed for organic no-till farming. Here, it is crushing a cover crop called hairy vetch into mulch, with little soil disturbance. This practice keeps most carbon in the ground, while the compressed crop suppresses weeds, contributes nitrogen and moisture to the soil, and creates habitat for insects beneficial to the corn that is simultaneously being planted.

There are several techniques for building up carbon in the soil. The first, and most commonly known, is preventing carbon from escaping the ground as it is tilled. Each time a plow turns over an acre of land, it releases an astounding 45 pounds of carbon. Organic farmers use a number of tillage systems to prevent carbon loss, including not tilling at all.

But the Rodale research has shown the viability of a second, and surprising, technique. Avoiding synthetic fertilizers and herbicides and keeping the soil covered with live plants builds the soil's organic content, which

help the soil retain carbon. Chemical fertilizers and weed killers essentially poison these fungi, hampering their carbon-storing ability.

In a world rapidly running out of time to reduce heat-trapping emissions, the promise of organic-style agriculture is welcome news. Implementing regenerative strategies for sequestering carbon requires no new technology or specialized knowledge. This suggests U.S. farmers, spurred by the right policies, could rapidly make the transition.

For more information, see *www.rodaleinstitute.org*.

CHAPTER 8
The Way Forward

No single solution is available to tackle global warming—the nation will need to enlist a full suite of policies and other incentives at the international, national, state, and local levels. Fundamentally, however, we need to shift to a clean energy future that can help solve three of our biggest challenges at once: breaking our dependence on oil, putting Americans back to work, and cutting carbon and other heat-trapping emissions to levels that will stave off some of the most devastating effects of global warming.

Fortunately, our analysis shows that it is technologically and economically feasible for the nation to achieve the needed cuts in emissions. In fact, the Blueprint also shows hundreds of billions of dollars of savings for consumers and businesses.

This chapter details some of the critical climate, energy, transportation, and international policies we need to address climate change. These policies form the building blocks of our clean energy future.

8.1. Building Block One: A Well-Designed Cap-and-Trade Policy

A central element of our climate policy should be a cap-and-trade system that sets tight limits on carbon emissions, and charges polluters for the emissions they do release. Legislation establishing such a system should require an auction through which industry

Federal legislation to reduce carbon emissions must be flexible enough to respond nimbly if new scientific information—such as fast-changing conditions in the Arctic—indicates the emissions cap or other measures should be strengthened. For this reason, federal legislation must include a science review provision, and the scientific recommendations must be evaluated and acted upon quickly.

Preserving tropical forests is an effective and fairly inexpensive way to curb a significant portion of the world's carbon emissions. Protecting these forests also benefits the people who depend on them—and the foods, products, and services the forests provide—for their lives and livelihoods, and preserves biological diversity. Federal legislation should allocate some revenue from the cap-and-trade system as an incentive to preserve tropical forests.

must purchase allowances to release those emissions (see Chapter 3).

That is an effective way to raise the revenues we need to invest in clean energy solutions; protect consumers, workers, and communities; and help people and wildlife adapt to the unavoidable effects of climate change. A cap-and-trade system that auctions allowances will also create a clear market signal that rewards cuts in carbon and other heat-trapping emissions and drives private investments in clean energy.

In designing an overall climate policy that includes cap and trade, U.S. policy makers must focus on several critical features:

Ensuring deep reductions in emissions. The United States must cut its total emissions at least 80 percent by 2050, and start on a path to achieving that goal by cutting emissions aggressively in the next 10 years. Government should set specific limits on carbon emissions from as many sources as possible, and provide incentives to cut emissions from other sources, to ensure that reductions will occur economywide. Our Blueprint analysis shows that the nation can meet a cap set at 26 percent below 2005 levels by 2020, and 56 percent below 2005 levels in

2030—taking us a considerable way toward meeting the 2050 target.

Rapidly responding to the latest science. Recent research is helping us understand how quickly and intensely the nation and the world are already feeling the effects of global warming. Any comprehensive response should therefore include a continuing review of the underlying science, and of the effectiveness of the U.S. program for addressing climate change. That approach should also be able to respond nimbly to the latest scientific information by setting new limits on emissions and creating new or more effective responses.

Funding protection for tropical forests in developing countries. Because tropical deforestation and forest degradation in developing nations contribute about 20 percent of worldwide global warming emissions, maintaining tropical forests is one of the most effective and least expensive ways to address global warming. A strong U.S. approach should channel a modest amount of revenue from the auction of carbon allowances to countries that preserve their forests, and also allow U.S. businesses subject to a cap on emissions to pay directly for a small number of carbon offsets in those countries.

Investing auction revenues wisely. As noted, government should auction carbon allowances and invest the revenue in programs and technologies that will help the nation shift to cleaner and more efficient energy. Government can also use auction revenues to help consumers pay energy bills and move to cleaner forms of energy and transportation, and provide transition assistance and job retention for workers and communities. Government can also use the funds to help U.S. companies remain globally competitive; help states, municipalities, tribes, and developing nations respond and adapt to the effects of global warming; and preserve threatened wildlife and ecosystems.

Containing costs appropriately. The most cost-effective way to tackle global warming is to invest heavily in energy efficiency measures, clean vehicles, and better transportation choices—all of which will drive down energy costs for consumers, businesses, municipalities, and states.

To enable companies subject to a cap on emissions to find the lowest-cost source of emissions cuts, a cap-and-trade system should allow such companies to purchase a limited number of carbon offsets: investments in reducing emissions from uncapped sectors, such as by paying farmers to adopt practices that allow soil to store more carbon.

How We Can Cut Emissions More than One-Third by 2020

Chapter 1 lays out a rationale for making significant cuts in U.S. carbon emissions by 2020, based on the urgency of the science and the need for a clear policy direction to move the nation toward a clean energy economy without delay.

Our findings show that the United States can cut global warming emissions 30 percent below 2005 levels (equivalent to 19 percent below 1990 levels) by 2020,* while providing substantial cost savings for consumers and businesses. And those figures do not include the full potential for storing carbon in the domestic agriculture and forest sectors. This is therefore a conservative estimate of the reductions that the nation could achieve domestically.

A separate UCS analysis shows that if our nation uses a modest amount of revenues from the auction of carbon allowances to help tropical nations reduce deforestation and forest degradation, the United States can reduce global warming emissions another 10 percent below 2005 levels (Boucher 2008). Negotiations on a global climate treaty now under way clearly show that the United States has the capacity and responsibility to finance even further reductions in carbon emissions by investing in the use of clean technology in developing countries. While these negotiations are still a work in progress, such investments could credit the United States with more cuts in emissions under a treaty.

Given the urgency of the science; the large potential for deep, cost-effective cuts revealed by the Blueprint; the danger that we will lock ourselves into high-carbon technologies; and the importance of meeting our global obligations, we recommend that the nation reduce emissions at least 35 percent below 2005 levels (or 25 percent below 1990 levels) by 2020, primarily through domestic action.

---

* Our modeling results show that capped firms over-comply with the cap set at 26 percent below 2005 levels in 2020, achieving actual reductions of 30 percent below 2005 levels.

Given the urgency of the science, the most expensive thing we can do is nothing. Recent scientific research suggests the effects of global warming are happening faster and more intensely than projected in the 2007 IPCC report. Coastal communities in the U.S. Northeast, where sea level is rising considerably faster and higher than the global mean, are particularly vulnerable. The Climate 2030 Blueprint lays out a clear and bold plan to curb carbon emissions.

However, such offsets must be limited, because firms in the capped sectors must have an incentive to alter their production and investment decisions if we are to meet our goals for sharply cutting global warming emissions and transition to cleaner technologies. Quality standards for offsets must also be closely monitored and enforced, so as not to compromise the nation's goals for cutting emissions.

Containing the costs of capped companies by creating a "safety valve"—setting an upper limit on the price of carbon allowances—would be unacceptable, because cuts in emissions could easily grind to a halt

By incorporating efficient design features and building materials, the LEED Platinum-certified Genzyme Center in Cambridge, MA, reduces its energy costs by more than 40 percent compared with a comparable building. Heat-retaining concrete slab construction moderates the interior temperature, while a double-paned glass curtain wall maximizes insulation and allows most employees to work in naturally lit space. Genzyme benefits from reduced energy costs while employees enjoy the improved quality of their work environment.

under such a policy and undermine the nation's entire effort to address climate change.

Preserving states' rights. Any policy should preserve rather than preempt the ability of states to implement their own more stringent climate, energy, and transportation policies.

8.2. Building Block Two: More Efficient Industries and Buildings

Making our industries and buildings more efficient must be a cornerstone of any comprehensive strategy for cutting carbon emissions. Energy efficiency can yield quick, significant, and sustained reductions in energy use, while providing substantial savings on energy bills for consumers and businesses. Creating a highly energy-efficient economy, however, requires policies and programs to help overcome significant and entrenched market barriers. The following policies build on the most effective approaches pursued by pioneering states and the federal government.

Enact an energy efficiency resource standard (EERS). Such a standard would require electricity and natural gas providers to meet targets for reducing their customers' energy use. It would also create a nationwide trading system for efficiency while spurring utilities to increase investments in efficiency. Some 18 states and countries such as France, Italy, and the United Kingdom have adopted such a standard.

Set new and higher energy efficiency standards for a broad range of appliances and equipment. Appliance and equipment standards save energy by requiring that various new products achieve minimum levels of efficiency by a certain date. Such standards have been one of the federal government's most successful strategies for reducing energy consumption in homes and businesses since their inception more than two decades ago.

Adopt more stringent energy efficiency codes for buildings. Stepping up energy codes over time ensures that builders deploy the most cost-effective technologies and best practices in all new residential and commercial construction.

Advance the deployment of combined-heat-and-power (CHP) systems. The nation can accomplish this by setting federal standards for permitting CHP systems and connecting them to the local power grid, and by establishing equitable interconnection fees and tariffs for standby, supplemental, and buy-back power. Greater funding for federal and state programs that spur the use of CHP through education, coordination, and direct project support is also needed.

8.3. Building Block Three: A Clean Future for Electricity

Energy experts have identified dozens of actions that policy makers can take now to reduce carbon emissions from the electricity sector. Here are a few.

Support a strong federal renewable electricity standard. Congress should enact a national standard requiring electric utilities to obtain at least 25 percent of their power from clean renewable sources by 2025. Studies have shown that such an approach is both feasible and affordable. Indeed, 28 states and the District of Columbia have adopted such standards, while the Senate has passed legislation establishing a standard three times, and the House of Representatives once. The national trading system in these bills would allow utilities to reduce their carbon emissions at an affordable price while creating jobs and stabilizing fuel prices.

Extend tax and other financial incentives for renewable energy. On-again/off-again extensions of tax credits for renewable energy have produced a boom-and-bust cycle that injects needless uncertainty into the financing and construction of such projects and raises their cost. Congress should also reduce incentives for fossil fuels and nuclear power, which are mature technologies that have already received enormous subsidies.

Triple today's federal funding for research and development of energy efficiency and renewable energy. A significant increase in R&D funding for clean energy technologies is needed to lower their costs and spur the widespread use essential to achieving dramatic cuts in carbon emissions.

Resolve state and local conflicts around siting electricity transmission lines and renewable energy projects. Policy makers must also reduce the state-by-state balkanization that is crippling creation of a nationwide grid for renewables.

New federal rules need to streamline siting efforts on federal lands while preserving their unique attributes and habitats. Congress should give the Federal Energy Regulatory Commission (FERC) more freedom to expedite new transmission capacity for renewable energy projects at the regional level. Congress also needs to reexamine provisions in the Federal Power Act that prohibit FERC from discriminating among power sources.

8.4. Building Block Four: A Smarter, Cleaner Transportation System

The transportation sector offers significant opportunities for cutting carbon emissions while reducing the cost of meeting our critical targets for addressing global warming. These reductions come from switching to low-carbon fuels and reducing our dependence on oil—which would also reduce consumers' and businesses' projected annual transportation costs about $120 billion by 2030. To achieve those cost savings, policy makers should create tools to strengthen each leg of the transportation sector: vehicles, fuels, and transportation choices.

Require investments in cleaner vehicles through tougher standards. The nation can save money and oil while cutting heat-trapping gases by

A strong federal renewable electricity standard—requiring utilities to obtain a percentage of their energy from renewable sources—will rapidly push these technologies into the marketplace. Twenty-eight states and the District of Columbia have already adopted such policies; not coincidentally, more wind power was installed in the United States over the past two years than in the previous 20. Wind represented 42 percent of all new electricity generating capacity installed in the country in 2007 and 2008.

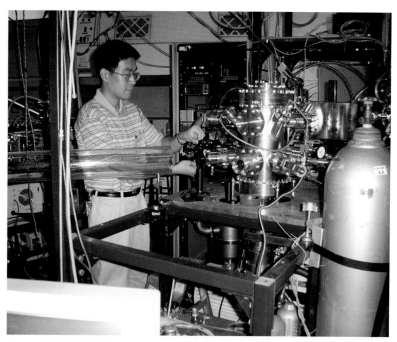

Research and development on emerging efficiency and renewable energy technologies—such as this researcher testing advanced photovoltaic equipment—should be increased. While implementing existing near-term solutions, we must continue to develop innovative technologies to ensure we achieve emissions reductions of at least 80 percent by 2050.

requiring automakers to integrate advanced technologies that boost fuel economy and reduce emissions from refrigerants across their entire fleet. Requiring cleaner, more efficient vehicles will also create jobs, help put the auto industry on the road to recovery, and ensure wise investment of public dollars used to help automakers.

Because many of these technologies fall under both the Clean Air Act, administered by the Environmental Protection Agency (EPA), and laws governing fuel economy, overseen by the National Highway Traffic Safety Administration (NHTSA), those two agencies can work together to set tougher standards for cars and light-duty trucks.

For example, the EPA should cap vehicle emissions from cars and light trucks at no more than 200 grams per mile of CO_2 equivalent by 2020, (with car and light-truck fuel economy reaching about 42 mpg), while NHTSA sets fuel-economy standards to support the EPA's efforts. By 2030 the EPA should cap vehicle emissions at no more than 140 grams per mile (with car and light-truck fuel economy reaching about 55 mpg). Within this process, there should be a transition to the EPA as the lead agency creating standards for vehicles in consultation with NHTSA.

In tackling medium- and heavy-duty vehicles, the EPA may be able to move more quickly than NHTSA,

as it has fewer restrictions on its statutory authority. The EPA's experience with setting standards for smog-producing and toxic emissions from heavy-duty vehicles, and its voluntary SmartWay fuel-saving program for such vehicles, should also prove valuable in the standard-setting process. Standards for medium-duty trucks should cap carbon emissions at no more than 780 grams per mile by 2020, and 500 grams per mile by 2030. Heavy-duty vehicles should emit no more than 1,075 grams per mile by 2020, and 840 grams per mile by 2030.

The EPA should set standards for all vehicles, not just highway vehicles, including airplanes, ships, off-road vehicles, and rail. All contribute to global warming, and all need to improve.

Require investments in cleaner fuels through a low-carbon fuel standard. The EPA also has an important role to play when it comes to fuels. A low-carbon fuel standard (LCFS)—which requires cuts in life-cycle carbon emissions per unit of energy delivered—is the next step up from today's renewable fuel standard (RFS).

The RFS applies to only about 10 percent of the transportation fuel pool, while an LCFS would encourage cuts in the carbon content of transportation fuels across the board. The latter would also avoid giving particular types of fuel special treatment, and allow the industry to determine the most cost-effective route to compliance.

The EPA already has authority under the Clean Air Act to establish a low-carbon fuel standard. The targets should be a 3.5 percent reduction in life-cycle carbon emissions from transportation fuels by 2020, and a 10 percent reduction by 2030. An LCFS would prevent an increase in global warming emissions from the use of high-carbon fuels such as tar sands, liquid coal, and oil shale. It would also guard against the types of biomass resources that could have that effect by spurring significant changes in land use. For an LCFS to be effective, it must take into account the full life cycle of a fuel, including both land-use changes and offshore emissions.

Maintain states' authority to set standards on global warming emissions from both vehicles and fuels. California's efforts to clean up smog and toxic pollution from vehicles, and encourage stronger sales of hybrid and electric vehicles, testify to the ability of states to act as laboratories for innovative energy and environmental policies. The next opportunity lies in California's efforts to reduce carbon emissions from cars, trucks, and fuels.

Congress must protect states' authority to develop such innovative policies and address new challenges as they emerge. That authority sustains progress when the federal government does not act quickly or aggressively enough, and it must be protected even as federal agencies establish national standards.

Encourage smarter travel, and include transportation under the carbon cap. Vehicles and fuels are just two parts of the transportation puzzle. To capture the remainder, a cap-and-trade system must include transportation. Doing so will send a price signal to all transportation users to reduce carbon emissions by choosing the best mode of transportation and curbing demand. Both pieces are crucial to meeting transportation's portion of the global warming challenge.

Besides including transportation under the cap, the federal government should tie all federal funding for transportation projects to efforts to cut carbon emissions. That will encourage innovative planning, improved mass transit, and intelligent transportation systems that make travel easier while reducing the need for it.

Federal agencies also need to encourage states to adopt pay-as-you-drive insurance, shift gas taxes to per-mile fees to sustain and expand revenues for repairing highways and expanding transit, and reward innovative local planning that encourages smarter growth and transportation options. Meanwhile states and localities must do their part to encourage alternatives to cars and trucks without sacrificing daily mobility, such as by making cities and towns more bike-friendly and walkable.

Encourage and invest in advanced transportation technologies. Federal support is also essential in developing, demonstrating, and deploying ultra-low-carbon vehicles, fuels, and infrastructure. The federal effort should focus on technologies that offer significant cuts in carbon emissions but that will have trouble entering the market on their own, such as low-carbon biofuels and vehicles that run on electricity or hydrogen from renewable energy sources.

However, all aspects of advanced transportation technologies need further R&D, from the basic science of batteries, fuel cells, and low-carbon biofuels to their low-cost manufacture and the infrastructure needed to sustain them. The federal government's role is especially critical given that the industry's investment in R&D is now in doubt due to severe financial challenges.

California has consistently led the way in efforts to clean up smog, encourage sales of electric vehicles, and, most recently, require carbon emissions reductions from vehicles and fuels. These policies should be adopted at the national level while preserving states' authority to push for cleaner vehicles and fuels.

To effectively address climate change, the United States must make deep cuts in its carbon emissions. However, there is also an urgent need to help reduce emissions in developing countries and to help poorer countries prepare for the changes we can no longer avoid. Comprehensive policies both domestically and abroad should help make renewable energy technologies, such as these PV modules installed on the roof of the Satyanarayanpur Health Center in West Bengal, India, more accessible and affordable.

Ensure that transportation policies are consistent and durable. The automotive industry needs certainty when making significant new investments, and the nation needs deep cuts in carbon emissions, so policies that encourage those investments and deliver those reductions must be strong and remain so even with a changing of the political guard. That is especially true for vehicle technologies, because 15 years can elapse before they exert their full impact as the fleet of cars and trucks turns over.

By consistently investing in a wide range of advanced technologies during the next 20 years rather than shifting focus with every new election or trend, the nation can ensure that we will have the tools we need to meet our transportation goals.

8.5. Building Block Five: International Policies

We were unable to model international policies in our analysis. However, we know that serious action to fight global warming will require the cooperation of all nations, as well as specific actions by industrialized countries. While the most important step our nation can take is to dramatically cut its own emissions, there is also an urgent need to help developing countries reduce their emissions and adapt to climate change. As a first step, the United States should engage constructively in U.N. negotiations now under way on a new climate treaty that keeps further warming below 2°F.

A comprehensive U.S. approach to global warming should include the following international policies.

Support for curbing tropical deforestation. Tropical deforestation now accounts for about 20 percent of heat-trapping emissions worldwide. Besides cutting back on its own emissions, the United States should finance and support the efforts of forest-rich tropical countries to slow their deforestation rates. A portion of the revenues from the auction of carbon allowances could fund this initiative. Investing just 5 percent of allowance revenues in this effort could reduce tropical deforestation by 20 percent (Boucher 2008).

Funding for sharing clean technology. Transitioning the global economy from its dependence on dirty fossil fuels to clean technologies will require serious investments in research, development, and wide-scale deployment. The United States should invest a portion of its auction revenues in efforts to share clean energy technologies, and should also consider agreements on intellectual property that would allow those technologies to be widely deployed more quickly.

Funding for adapting to global warming. Unfortunately, the world is already committed to a certain amount of global warming because of past and current carbon emissions. Particularly vulnerable communities and regions are already experiencing the effects of climate change, and will continue to bear the brunt. The United States and other developed nations must fund efforts to help these communities and regions build

BOX 8.2.

How It Works: REDD

Policies for reducing emissions from deforestation in developing countries—known as REDD—can be very cost-effective ways to slow global warming. The opportunity costs of preserving tropical forest-land, which are the majority of the costs of REDD, are low because most deforested land is used in ways that bring very low returns. For example, 60–70 percent of Amazon land deforested in the 1990s was used for low-quality cattle pasture, with many acres required to support a single cow.

A UCS analysis (Boucher 2008) describes how REDD could work in practice. Some of the highlights of that analysis are described below.

Under a REDD system, developed nations would compensate tropical nations for these opportunity costs once the tropical nations had slowed their rates of deforestation and documented the resulting cuts in carbon emissions (calculated for each country as a whole). Funding could come from a variety of sources, such as auction revenues from cap-and-trade systems, official development assistance, or levies on aviation fuels or timber imports.

If funding for curbing deforestation came from carbon offsets purchased by companies in developed countries like the United States, net emissions would not drop, as a cut in emissions in the tropical country would be countered by more emissions in the United States. Our REDD modeling assumed that U.S. funding for REDD would come from a non-offset source such as auction revenues. As a basis for comparison, The European Commission recommends using 5 percent of auction revenues under the EU Emissions Trading System for this purpose.

Three major groups of researchers have modeled the costs and potential of REDD (Kindermann et al. 2008). Our analysis averaged the output of their models—modified to incorporate other costs of implementing a REDD program, and realistic expectations for how quickly it could become truly global—to create a new set of cost curves.

The analysis found costs of REDD that are comparable to those of other recent studies. For example, cutting tropical deforestation in half by 2020—the goal announced by both the U.K.'s Eliasch Review and the European Commission's October 2008 report on

REDD—would cost about $20 billion a year. The EC estimated an annual cost of $15 billion to $25 billion, while the Eliasch Review cited a range of $18 billion to $26 billion. Thus a variety of estimates of the cost of REDD are converging on the same relatively modest figures.

If funding were available, would tropical countries reduce their deforestation rates? Indications are that the answer is yes. About 30 members of the Coalition for Rain Forest Nations put REDD on the agenda of international climate talks in 2005. And several tropical countries have already stopped and even reversed deforestation (Rudel et al. 2005), along with most temperate ones.

For example, Brazil recently released its National Climate Change Plan, which aims to reduce deforestation by slightly more than 70 percent through 2017, compared with the baseline level from 1996 to 2005. (The nation will measure progress in hectares deforested rather than tons of CO_2 emitted, but the results should be similar.) Thus, if developed nations can find a relatively small amount of funding, developing countries seem willing and able to accomplish ambitious goals.

About 20 percent of global carbon emissions result from the destruction of tropical forests. A sensible global warming policy must contain financial incentives for developing countries to protect their forests.

resilience in the face of climate change, especially the poorest areas. A portion of the auction revenues from a cap-and-trade program could augment existing funding for international development.

8.6. Conclusion

We are at a crossroads. The Reference case shows that we are on a path of rising energy use and heat-trapping emissions. We are already seeing significant impacts from this carbon overload, such as rising temperatures and sea levels and extreme weather events. If carbon emissions continue to climb at their current rate, we could reach climate "tipping points" and face irreversible changes to our planet.

In 2007 the Intergovernmental Panel on Climate Change (IPC) found it "unequivocal" that Earth's climate is warming, and that human activities are the primary cause (IPCC 2007). The IPCC report concludes that unchecked global warming will only create more adverse impacts on food production, public health, and species survival.

The climate will not wait for us. More recent studies have shown that the measured impacts—such as rising sea levels and shrinking summer sea ice in the Arctic—are occurring more quickly, and often more intensely, than IPCC projections (Rosenzweig et al. 2008; Rahmstorf et al. 2007; Stroeve et al. 2007).

The most expensive thing we can do is nothing. One study estimates that if climate trends continue, the total cost of global warming in the United States could be as high as 3.6 percent of GDP by 2100 (Ackerman and Stanton 2008).

The Climate 2030 Blueprint demonstrates that we can choose to cut our carbon emissions while maintaining robust economic growth and achieving significant energy-related savings. While the Blueprint policies are not the only path forward, a near-term, comprehensive suite of climate, energy, and transportation policies is essential if we are to curb global warming in an economically sound fashion. These near-term policies are also only the beginning of the journey toward achieving a clean energy economy. The nation can and must expand these and other policies beyond 2030 to ensure that we meet the mid-century reductions in emissions that scientists deem necessary to avoid the worst consequences of global warming.

The Climate 2030 Blueprint demonstrates that we can cut our carbon emissions while maintaining robust economic growth and achieving significant energy-related savings. A near-term, comprehensive suite of climate, energy, and transportation policies is essential to cost-effectively curb global warming and build a revitalized clean energy economy.

FIGURE 8.1. Choosing a Clean Energy Economy

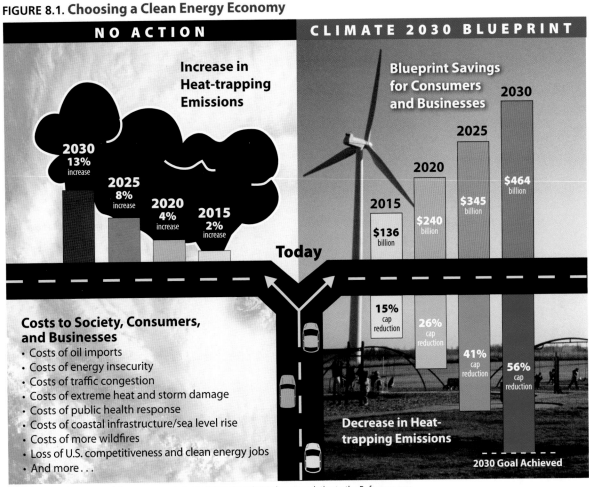

Note: Emissions increases and decreases are relative to 2005; Blueprint savings are relative to the Reference case.

The United States is at a crossroads. We can choose to transition to a clean energy economy that addresses a multitude of challenges (oil dependency, energy security, global warming, air pollution) or we can choose to ignore these problems.

The Climate 2030 Blueprint shows that we can build a competitive clean energy economy that will save consumers money and give our children a healthy future.

Conversely, choosing to ignore our energy problems commits us to continued reliance on dirty fossil fuels and to the damaging costs associated with climate change. These costs include the consequences of sea level rise that threaten our coastal communities, disruptions in food production, and illnesses associated with extreme heat and diminished air quality.

This transition will certainly require some up-front investment costs. However, the Climate 2030 Blueprint will reduce energy use and consumer and business energy bills—even in the early years. These savings more than make up for the costs of building a clean energy economy.

The time to invest in our future is now.

References

Executive Summary

Ackerman, F., and E.A. Stanton. 2008. *The cost of climate change: What we'll pay if global warming continues unchecked.* New York, NY: Natural Resources Defense Council. Online at *http://www.nrdc.org/globalWarming/cost/cost.pdf*.

Intergovernmental Panel on Climate Change (IPCC). 2007. *Climate change 2007: The physical science basis.* Contribution of Working Group I to the Fourth Assessment Report of the Intergovernmental Panel on Climate Change, edited by S. Solomon, D. Qin, M. Manning, Z. Chen, M. Marquis, K.B. Averyt, M.Tignor, and H.L. Miller. Cambridge, UK: Cambridge University Press.

Luers, A.L., M.D. Mastrandrea, K. Hayhoe, and P.C. Frumhoff. 2007. *How to avoid dangerous climate change: A target for U.S. emissions reductions.* Cambridge, MA: Union of Concerned Scientists.

Rahmstorf, S., A. Cazenave, J.A. Church, J.E. Hansen, R.F. Keeling, D.E. Parker, and R.C.J. Somerville. 2007. Recent climate observations compared to projections. *Science* 316:709.

Rosenzweig, C., D. Karoly, M. Vicarelli, P. Neofotis, Q. Wu, G. Casassa, A. Menzel, T.L. Root, N. Estrella, B. Seguin, P. Tryjanowski, C. Liu, S. Rawlins, and A. Imeson. 2008. Attributing physical and biological impacts to anthropogenic climate change. *Nature* 453:353–357.

Sperling, D., and D. Gordon. 2009. *Two billion cars: Driving toward sustainability.* New York: Oxford University Press.

Stroeve, J., M. Serreze, S. Drobot, S. Gearheard, M. Holland, J. Maslanik, W. Meier, and T. Scambos. 2008. Arctic sea ice extent plummets in 2007. *Eos, Transactions, American Geophysical Union* 89(2):13–20.

Chapter 1
A Vision of a Clean Energy Economy

American Council for Capital Formation (ACCF) and National Association of Manufacturers (NAM). 2008. *Analysis of the Lieberman-Warner Climate Security Act (S.2191) using the national energy modeling system.* Analysis conducted by Science Applications International Corporation. Washington, DC. Online at *http://www.accf.org/pdf/NAM/fullstudy031208.pdf*.

American Physical Society (APS). 2008. Energy future: Think efficiency. American Physical Society. Online at *http://www.aps.org/energyefficiencyreport/report/aps-energyreport.pdf*.

American Solar Energy Society (ASES). 2007. *Tacking climate change in the U.S.: Potential carbon emissions reductions from energy efficiency and renewable energy by 2030*, edited by C.F. Kutscher. Boulder, CO. Online at *http://www.ases.org/images/stories/file/ASES/climate_change.pdf*.

Baer, P., T. Athanasiou, S. Kartha, and E. Kemp-Benedict. 2008. *The greenhouse development rights framework: The right to development in a climate constrained world.* Revised second edition. Publication Series on Ecology. Berlin: The Heinrich Böll Foundation, Christian Aid, EcoEquity, and the Stockholm Environment Institute.

Baer, P., and M.D. Mastrandrea. 2006. High stakes: Designing emissions pathways to reduce the risk of dangerous climate change. Institute for Public Policy and Research.

Banks, J. 2008. The Lieberman-Warner Climate Security Act—S.2191: A summary of modeling results from the national energy modeling system. Boston, MA: Clean Air Task Force.

Climate Change Research Centre. 2007. *The 2007 Bali climate declaration by scientists.* Sydney: University of New South Wales. Online at *http://www.ccrc.unsw.edu.au/news/2007/Bali.html*, accessed April 6, 2009.

Congressional Budget Office (CBO). 2007. *The potential for carbon sequestration in the United States.* Publication no. 2931. Washington, DC.

den Elzen, M.G.J., N. Höhne, J. van Vliet, and C. Ellermann. 2008. *Exploring comparable post-2012 reduction efforts for Annex I countries.* Report 500102019/2008. Netherlands Environmental Assessment Agency (PBL), 92.

Electric Power Research Institute (EPRI). 2007. *The power to reduce CO_2 emissions: The full portfolio.* Discussion paper prepared for the EPRI 2007 Summer Seminar attendees. Palo Alto, CA.

Energy Information Administration (EIA). 2008. Energy market and economic impacts of S.2191, the Lieberman-Warner Climate Security Act of 2007. Washington, DC: U.S. Department of Energy.

Energy Information Administration (EIA). 2007. *Energy and economic impacts of implementing both a 25-percent renewable portfolio standard and a 25-percent renewable fuel standard by 2025.* Washington, DC: U.S. Department of Energy.

Environmental Protection Agency (EPA). 2008. *Inventory of U.S. greenhouse gas emissions and sinks: 1990–2006.* EPA 430-R-08-005. Washington, DC.

Environmental Protection Agency (EPA). 2008a. *EPA analysis of the Lieberman-Warner Climate Security Act of 2008.* Washington, DC. Online at *http://www.epa.gov/climatechange/downloads/s2191_EPA_Analysis.pdf.*

Flavin, C. 2008. *Low-carbon energy: A roadmap.* Washington, DC: Worldwatch Institute. Online at *http://www.worldwatch.org/node/5945.*

Google. 2008. Clean energy 2030: Google's proposal for reducing U.S. dependence on fossil fuels. Online at *http://www.google.com/energyplan*, accessed on April 13, 2009.

Greenpeace International and the European Renewable Energy Council. 2009. *Energy [r]evolution: A sustainable USA energy outlook.* Washington, DC. Online at *http://www.greenpeace.org/raw/content/usa/press-center/reports4/energy-r-evolution-a-sustain.pdf.*

Gupta, S., D.A. Tirpak, N. Burger, J. Gupta, N. Höhne, A.I. Boncheva, G.M. Kanoan, C. Kolstad, J.A. Kruger, A. Michaelowa, S. Murase, J. Pershing, T. Saijo, and A. Sari. 2007. Policies, instruments and co-operative arrangements. In *Climate change 2007: Mitigation.* Contribution of Working Group III to the Fourth Assessment Report of the Intergovernmental Panel on Climate Change, edited by B. Metz, O.R. Davidson, P.R. Bosch, R. Dave, and L.A. Meyer. Cambridge, UK: Cambridge University Press.

Hansen, J., M. Sato, P. Kharecha, D. Beerling, R. Berner, V. Masson-Delmotte, M. Pagani, M. Raymo, D.L. Royer, and J.C. Zachos. 2008. Target atmospheric CO_2: Where should humanity aim? *The Open Atmospheric Science Journal* 2:217–223.

Hansen, J., L. Nazarenko, R. Ruedy, M. Sato, J. Willis, A. Del Genio, D. Koch, A. Lacis, K. Lo, S. Menon, T. Novakov, J. Perlwitz, G. Russell, G.A. Schmidt, and N. Tausnev. 2005. Earth's energy imbalance: Confirmation and implications. *Science* 308:1431–1435.

Intergovernmental Panel on Climate Change (IPCC). 2007. *Climate change 2007: The physical science basis.* Contribution of Working Group I to the Fourth Assessment Report of the Intergovernmental Panel on Climate Change, edited by S. Solomon, D. Qin, M. Manning, Z. Chen, M. Marquis, K.B. Averyt, M.Tignor, and H.L. Miller. Cambridge, UK: Cambridge University Press.

Intergovernmental Panel on Climate Change (IPCC). 2007a. Summary for policymakers. In *Climate change 2007: Impacts, adaptation and vulnerability.* Contribution of Working Group II to the Fourth Assessment Report of the Intergovernmental Panel on Climate Change, edited by M.L. Parry, O.F. Canziani, J.P. Palutikof, P.J. van der Linden, and C.E. Hanson. Cambridge, UK: Cambridge University Press, 7–22.

Intergovernmental Panel on Climate Change (IPCC). 2007b. Summary for policymakers. In *Climate change 2007: Mitigation.* Contribution of Working Group III to the Fourth Assessment Report of the Intergovernmental Panel on Climate Change, edited by B. Metz, O.R. Davidson, P.R. Bosch, R. Dave, and L.A. Meyer. Cambridge, UK: Cambridge University Press.

Loulergue, L., A. Schilt, R. Spahni, V. Masson-Delmotte, T. Blunier, B. Lemieux, J.-M. Barnola, D. Raynaud, T.F. Stocker, and J. Chappellaz. 2008. Orbital and millennial-scale features of atmospheric CH_4 over the past 800,000 years. *Nature* 453:383–386.

Luers, A.L., M.D. Mastrandrea, K. Hayhoe, and P.C. Frumhoff. 2007. *How to avoid dangerous climate change: A target for U.S. emissions reductions.* Cambridge, MA: Union of Concerned Scientists.

Lüthi, D., M. Le Floch, B. Bereiter, T. Blunier, J.-M. Barnola, U. Siegenthaler, D. Raynaud, J. Jouzel, H. Fischer, K. Kawamura, and T.F. Stocker. 2008. High-resolution carbon dioxide concentration record 650,000–800,000 years before present. *Nature* 453:379–382.

McKinsey and Company. 2009. *Pathways to a low-carbon economy: Version 2 of the global greenhouse gas abatement cost curve.* New York.

Meehl, G.A., W.M. Washington, W.D. Collins, J.M. Arblaster, A. Hu, L.E. Buja, W.G. Strand, and H. Teng. 2005. How much more global warming and sea level rise? *Science* 307(5716):1769–1772.

Meinshausen, M., W.L. Hare, T.M.L. Wigley, D.P. van Vuuren, M.G.J. den Elzen, and R. Swart. 2006. Multi-gas emission pathways to meet climate targets. *Climatic Change* 75:151–194.

Murray, B.C., B. Sohngen, A.J. Sommer, B. Depro, K. Jones, B. McCarl, D. Gillig, B. DeAngelo, and K. Andrasko. 2005. *Greenhouse gas mitigation potential in U.S. forestry and agriculture.* EPA 430-R-05-006. Washington, DC: Environmental Protection Agency, Office of Atmospheric Programs.

Pacala, S., and R. Socolow. 2004. Stabilization wedges: Solving the climate problem for the next 50 years with current technologies. *Science* 305:968–971.

Paltsev, S., J.M. Reilly, H. Jacoby, A. Gurgel, G.E. Metcalf, A.P. Sokolov, and J.F. Holak. 2007. *Assessment of U.S. cap-and-trade proposals.* Report no. 146. Cambridge, MA: Massachusetts Institute of Technology Joint Program on the Science and Policy of Global Change.

Rahmstorf, S., A. Cazenave, J.A. Church, J.E. Hansen, R.F. Keeling, D.E. Parker, and R.C.J. Somerville. 2007. Recent climate observations compared to projections. *Science* 316:709.

Rosenzweig, C., D. Karoly, M. Vicarelli, P. Neofotis, Q. Wu, G. Casassa, A. Menzel, T.L. Root, N. Estrella, B. Seguin, P. Tryjanowski, C. Liu, S. Rawlins, and A. Imeson. 2008. Attributing physical and biological impacts to anthropogenic climate change. *Nature* 453:353–357.

Smith, J.B., S.H. Schneider, M. Oppenheimer, G.W. Yohe, W. Hare, M.D. Mastrandrea, A. Patwardhan, I. Burton, J. Corfee-Morlot, C.H.D. Magadza, H.-M. Füssel, A.B. Pittock, A. Rahman, A. Suarez, and J.-P. van Ypersele. 2009. Assessing dangerous climate change through an update of the Intergovernmental Panel on Climate Change (IPCC) reasons for concern. *Proceedings of the National Academy of Sciences* doi:10.1073/pnas.0812355106. Online at *http://www.pnas. org/content/early/2009/02/25/0812355106*.

Smith, J.B., H.J. Schellnhuber, and M.M. Qader Mirza. 2001. Vulnerability to climate change and reasons for concern: A synthesis. In *Climate change 2001: Impacts, adaptation and vulnerability*, edited by J. McCarthy, O. Canziana, N. Leary, D. Dokken, and K. White. New York: Cambridge University Press, 913–967. Online at *http://www.grida.no/ climate/ipcc_tar/wg2/pdf/wg2TARchap19.pdf*.

Stroeve, J., M. Serreze, S. Drobot, S. Gearheard, M. Holland, J. Maslanik, W. Meier, and T. Scambos. 2008. Arctic sea ice extent plummets in 2007. *Eos, Transactions, American Geophysical Union* 89(2):13–20.

Tans, P. 2009. *Trends in atmospheric carbon dioxide.* Boulder, CO: National Oceanic and Atmospheric Administration Earth System Research Laboratory. Online at *http://www.esrl. noaa.gov/gmd/ccgg/trends*, accessed on April 6, 2009.

Union of Concerned Scientists (UCS). 2008. *U.S. scientists and economists' call for swift and deep cuts in greenhouse gas emissions.* Cambridge, MA.

van Vuuren, D.P., M.G.J. den Elzen, P.L. Lucas, B. Eickhout, B.J. Strengers, B. van Ruijven, S. Wonink, and R. van Houdt. 2007. Stabilizing greenhouse gas concentrations at low levels: An assessment of reduction strategies and costs. *Climatic Change* 81:119–159.

Wigley, T.M.L. 2005. The climate change commitment. *Science* 307:1766–1769.

Success Story: Reinventing Pittsburgh as a Green City

City of Pittsburgh. 2008. Mayor partners with the Pittsburgh project to green up abandoned baseball field. Press release. Online at *http://www.city.pittsburgh.pa.us/mayor/photos_ ravenstahl/2008/june.html#June_18,_2008_Green_Trust*, accessed on February 4, 2009.

City of Pittsburgh. 2007. Ordinance supplementing the Pittsburgh Code, Title Nine, Zoning, by adding a new sub section 915.06, entitled, "Sustainable development for publicly financed buildings." Legislative file number 2007–1950. Online at *http://legistar.city.pittsburgh.pa.us/detailreport/ Reports/Temp/4292009152213.pdf*.

David L. Lawrence Convention Center (DLCC). 2009. Being green. Pittsburgh, PA. Online at *http://www.pittsburghcc.com/ cc/_pdf/BeingGreen.pdf*, accessed on January 29, 2009.

Global Insight Inc. 2008. U.S. metro economies: Current and potential jobs in the U.S. economy. Prepared for the U.S. Conference of Mayors and the Mayors Climate Protection Center. Online at *http://www.usmayors.org/pressreleases/uploads/ GreenJobsReport.pdf*, accessed on April 3, 2009.

Green Building Council (USGBC). 2008. LEED benchmarking data: July 2008. Washington, DC. Online at *http://www. gbapgh.org/Files/July08LEEDBenchmarkingData.pdf*, accessed on March 31, 2009.

Malone, R. 2007. Which are the world's cleanest cities? *Forbes*, April 16.

Office of Energy Efficiency and Renewable Energy (EERE). 2009. *Building energy codes program: PNC Firstside Center.* Washington, DC: U.S. Department of Energy. Online at *http://www.energycodes.gov/news/ecodes2007/presentations/pre/ tours_PNC_Firstside_Center.pdf*.

Parton, J. 1868. Pittsburg. *The Atlantic Monthly*, January. Online at *http://www.pittsburghinwords.org/james_parton.html*.

The Pittsburgh Channel. 2008. Blue Green alliance uses PNC Firstside Center as environmentally friendly example. August 8. Online at *http://www.thepittsburghchannel.com/green-pages/ 17136311/detail.html*, accessed on February 4, 2009.

Plextronics. 2009. Plextronics marks the opening of its new development line at a ribbon-cutting ceremony. Pittsburgh, PA. Online at *http://www.plextronics.com/press_detail. aspx?PressReleaseID=95*, accessed on February 4, 2009.

Schooley, T. 2008. Wind power company plans Pittsburgh location. *Pittsburgh Business Times*, August 21.

Solar Power Industries. 2009. Solar power industries, solar cell manufacturer. Belle Vernon, PA. Online at *http://www. solarpowerindustries.com/index.html*, accessed on February 4, 2009.

Sports and Exhibition Authority (SEA). 2008. David L. Lawrence convention center. Pittsburgh, PA. Online at *http:// www.pgh-sea.com/conventioncenter.htm*, accessed on January 29, 2009.

University of Pittsburgh. 2009. Environmental innovations in Pittsburgh. Pittsburgh, PA. Online at *http://www.pitt.edu/ ~esweb*, accessed on February 4, 2009.

Chapter 2
Our Approach

American Association of State and Highway Transportation Officials (AASHTO). 2007. *Transportation: Invest in our future.* Washington, DC. Online at *http://www.transportation1.org/tif5report/TIF5.pdf.*

Bandivadekar, A., K. Bodek, L. Cheah, C. Evans, T. Groode, J. Heywood, E. Kasseris, M. Kromer, and M. Weiss. 2008. *On the road in 2035: Reducing transportation's petroleum consumption and GHG emissions.* LFEE 2008-05 RP. Cambridge, MA: Massachusetts Institute of Technology Laboratory for Energy and the Environment.

Bedsworth, L.W. 2004. *Climate control: Global warming solutions for California cars.* Cambridge, MA: Union of Concerned Scientists. Online at *http://www.ucsusa.org/assets/documents/clean_vehicles/climate_control_-lo-res.pdf.*

Bordoff, J.E., and P.J. Noel. 2008. Pay-as-you-drive auto insurance: A simple way to reduce driving-related harms and increase equity. Washington, DC: The Brookings Institution, July. Online at *http://www.brookings.edu/papers/2008/07_payd_bordoffnoel.aspx,* accessed on February 20, 2009.

California Air Resources Board (CARB). 2008. *Comparison of greenhouse gas reductions for the United States and Canada under U.S. CAFE standards and California Air Resources Board greenhouse gas regulations: An enhanced technical assessment.* Online at *http://www.climatechange.ca.gov/publications/arb/ARB-1000-2008-012/ARB-1000-2008-012.PDF.*

Cooper, C., R. Reinhart, M. Kromer, R. Wilson, and R. Schubert. Forthcoming. *Reducing heavy-duty vehicle fuel consumption and GHG emissions,* edited by P. Miller. Boston, MA: Northeast States Center for a Clean Air Future (NESCCAF); San Francisco, CA: International Council on Clean Transportation (ICCT).

Cowart, W. 2008. How much can we slow VMT growth? Presented to CCAP VMT & Climate Change Workshop. Cambridge, MA: Cambridge Systematics. Online at *http://www.ccap.org/docs/resources/460/How%20much%20can%20we%20slow%20VMT%20growth%20May%202008.pdf.*

Energy Information Administration (EIA). 2009. *Annual energy outlook 2009 with projections to 2030.* Washington, DC: U.S. Department of Energy. Online at *http://www.eia.doe.gov/oiaf/aeo/aeoref_tab.html.*

Energy Information Administration (EIA). 2008. Energy market and economic impacts of S.2191, the Lieberman-Warner Climate Security Act of 2007. Washington, DC: U.S. Department of Energy.

Energy Information Administration (EIA). 2008a. *Annual energy outlook 2008 with projection to 2030.* Washington, DC: U.S. Department of Energy.

Energy Information Administration (EIA). 2007. *Energy and economic impacts of implementing both a 25-percent renewable portfolio standard and a 25-percent renewable fuel standard by 2025.* Washington, DC: U.S. Department of Energy.

Freese, B., S. Clemmer, and A. Nogee. 2008. *Coal power in a warming world: A sensible transition to cleaner energy options.* Cambridge, MA: Union of Concerned Scientists. Online at *http://www.ucsusa.org/assets/documents/clean_energy/Coal-power-in-a-warming-world.pdf.*

Interlaboratory Working Group (IWG). 2000. *Scenarios for a clean energy future.* ORNL/CON-476 and LBNL-44029. Oak Ridge, TN: Oak Ridge National Laboratory; and Berkeley, CA: Lawrence Berkeley National Laboratory.

National Research Council (NRC). 2002. *Effectiveness and impact of Corporate Average Fuel Economy (CAFE) standards.* The National Academies Press.

Office of Energy Efficiency and Renewable Energy (EERE). 2008. *20% wind energy by 2030: Increasing wind energy's contribution to U.S. electricity supply.* DOE/GO-102008-2578. Washington, DC: U.S. Department of Energy. Online at *http://www.20percentwind.org/20p.aspx?page=Report.*

Chapter 3
Putting a Price on Global Warming Emissions

City of Chicago. 2008. *Chicago climate action plan,* 15. Online at *http://www.chicagoclimateaction.org/filebin/pdf/finalreport/CCAPREPORTFINAL.pdf.*

Congressional Budget Office (CBO). 2007a. *Tradeoffs in allocating allowances for CO_2 emissions.* Economic and budget issue brief, April 25. Washington, DC.

Dinan, T.M., and D.L. Rogers. 2002. Distribution effects of carbon allowance trading: How government decisions determine winners and losers. *National Tax Journal* 55(2), June.

Environmental Protection Agency (EPA). 2007. *Acid rain and related programs.* 2007 progress report. Washington, DC. Online at *http://www.epa.gov/airmarkets/progress/docs/2007ARPReport.pdf.*

Luers, A.L., M.D. Mastrandrea, K. Hayhoe, and P.C. Frumhoff. 2007. *How to avoid dangerous climate change: A target for U.S. emissions reductions.* Cambridge, MA: Union of Concerned Scientists.

Stern, N. 2006. The Stern review: The economics of climate change. London: HM Treasury, Office of Climate Change. Online at *http://www.hm-treasury.gov.uk/sternreview_index.htm.*

Chapter 4
Where We Work, Live and Play

Amann, J.T., A. Wilson, and K. Ackerly. 2007. *Consumer guide to home energy savings*. Ninth edition. Washington, DC: American Council for an Energy-Efficient Economy.

American Council for an Energy-Efficient Economy (ACEEE). 2007. *2007 federal energy legislation: Estimates of energy and carbon savings from energy bill passed in Senate 12/14/2007*. Washington, DC. Online at *http://www.aceee. org/energy/national/07nrgleg.htm.*

American Solar Energy Society (ASES). 2007. *Tacking climate change in the U.S.: Potential carbon emissions reductions from energy efficiency and renewable energy by 2030*, edited by C.F. Kutscher. Boulder, CO. Online at *http://www.ases.org/images/ stories/file/ASES/climate_change.pdf.*

Brooks, S., B. Elswick, and R.N. Elliott. 2006. *Combined heat and power: Connecting the gap between markets and utility interconnection and tariff practices (Part I)*. Washington, DC: American Council for an Energy-Efficient Economy. Online at *http://aceee.org/pubs/ie062.htm.*

Brown, E., R.N. Elliott, and S. Nadel. 2005. *Energy efficiency programs in agriculture: Design, success, and lessons learned*. Washington, DC: American Council for an Energy-Efficient Economy. Online at *http://aceee.org/pubs/ie051.htm.*

California Energy Commission (CEC). 2007. *Integrated energy policy report*. CEC-100-2007-008-CMF. Sacramento, CA. Online at *http://www.energy.ca.gov/2007_energypolicy/ documents/index.html.*

Creyts, J., A. Derkach, S. Nyquist, K. Ostrowski, and J. Stephenson. 2007. *Reducing U.S. greenhouse gas emissions: How much at what cost?* New York: McKinsey and Company.

Efficiency Vermont. 2007. Annual report. Burlington, VT. Online at *http://www.efficiencyvermont.com.*

Ehrhardt-Martinez, K., and J.A. Laitner. 2008. *The size of the U.S. energy efficiency market: Generating a more complete picture*. ACEEE report E083. Washington, DC: American Council for an Energy-Efficient Economy.

Elliott, R.N., M. Eldridge, A.M. Shipley, J. Laitner, S. Nadel, A. Silverstein, B. Hedman, and M. Sloan. 2007a. *Potential for energy efficiency, demand response and onsite renewable energy to meet Texas' growing electricity demands*. ACEEE report E073. Washington, DC: American Council for an Energy-Efficient Economy.

Elliott, R.N., M. Eldridge, A.M. Shipley, J. Laitner, S. Nadel, P. Fairey, R. Vieira, J. Sonne, A. Silverstein, B. Hedman, and K. Darrow. 2007b. *Potential for energy efficiency and renewable energy to meet Florida's growing energy demands*, ACEEE Report E072. Washington, DC.: American Council for an Energy-Efficient Economy.

Elliott, R.N., T. Langer, and S. Nadel. 2006. *Reducing oil use through energy efficiency: Opportunities beyond light cars and trucks*. ACEEE report E061. Washington, DC: American Council for an Energy-Efficient Economy.

Energy Information Administration (EIA). 2009. *Annual energy outlook 2009 with projections to 2030*. Washington, DC: U.S.Department of Energy. Online at *http://www.eia. doe.gov/oiaf/aeo/aeoref_tab.html.*

Energy Information Administration (EIA). 2008a. *Annual energy outlook 2008 with projection to 2030*. Washington, DC: U.S. Department of Energy.

Energy Information Administration (EIA). 2008b. *2003 commercial buildings energy consumption survey*. Table E1A. Washington, DC: U.S. Department of Energy.

Energy Information Administration (EIA). 2005. *2005 residential energy consumption survey*. Table US12. Washington, DC: U.S. Department of Energy.

Energy Information Administration (EIA). 2000. *The market and technical potential for combined heat and power in the industrial sector*. Washington, DC: U.S. Department of Energy.

Environmental Protection Agency (EPA). 2008b. High efficiency water heaters. Washington, DC. Online at *http://www. energystar.gov/ia/new_homes/features/WaterHtrs_062906.pdf*, accessed on January 16, 2009.

Global Energy Partners. 2003. *California summary study of 2001 energy efficiency programs: Final report*. Lafayette, CA.

Hurley, D., K. Takahashi, B. Biewald, J. Kallay, and R. Maslowski. 2008. *Costs and benefits of electric utility energy efficiency in Massachusetts*. Prepared for the Northeast Energy Efficiency Council. Cambridge, MA: Synapse Energy Economics.

Interlaboratory Working Group (IWG). 2000. *Scenarios for a clean energy future*. ORNL/CON-476 and LBNL-44029. Oak Ridge, TN: Oak Ridge National Laboratory; and Berkeley, CA: Lawrence Berkeley National Laboratory.

Laiter, J.A. 2008. Working calculations based on: Energy Information Administration (EIA). 2008. *Short term energy outlook*. U.S. Department of Energy. And: Energy Information Administration (EIA). 2008. *Annual energy outlook 2008 with projections to 2030*. U.S. Department of Energy. And: Energy Information Administration (EIA). 2006. *Annual energy outlook 2006 with projections to 2030*. U.S. Department of Energy. Washington, DC: American Council for an Energy-Efficient Economy.

Nadel, S. 2007. *Energy efficiency resource standards around the U.S. and the world*. Washington, DC: American Council for an Energy-Efficient Economy.

Nadel, S. 2006. *Energy efficiency resource standards: Experience and recommendations*. ACEEE report E063. Washington, DC: American Council for an Energy-Efficient Economy.

Nadel, S., A. deLaski, M. Eldridge, and J. Kliesch. 2006. *Leading the way: Continued opportunities for new state appliance and equipment efficiency standards.* ACEEE report A061. Washington, DC: American Council for an Energy-Efficient Economy.

Nadel, S., A. Shipley, and R.N. Elliott. 2004. *The technical and achievable potential for energy efficiency in the U.S.: A meta-analysis of recent studies.* A proceeding of the 2004 ACEEE summer study on energy efficiency in buildings. Washington, DC: American Council for an Energy-Efficient Economy.

Office of Energy Efficiency and Renewable Energy (EERE). 2006. *Walls and roofs.* Washington, DC: U.S. Department of Energy. Online at *www.eere.energy.gov/states/alternatives/walls_roofs.cfm.*

Ondrey, G. 2004. Controlling the enterprise: The next generation of automation systems move beyond process control and target the best return on assets. *Chemical Engineering* 111(11):19.

President's Committee of Advisors on Science and Technology (PCAST). 1997. *Federal energy research and development for the challenges for the twenty-first century.* Report of the Energy Research and Development Panel. Washington, DC: Office of Science and Technology Policy.

Prindle, W., J. Sathaye, S. Murtishaw, D. Crossley, G. Watt, J. Hughes, M. Takahashi, H. Asano, E. Worrell, S. Joosen, E. de Visser, M. Harmelink, J. Bjorndalen, J. Bugge, and B. Baardson. 2007. *Quantifying the effects of market failures in the end-use of energy.* Washington, DC: American Council for an Energy-Efficient Economy.

Sachs, H., S. Nadel, J.T. Amann, M. Tuazon, E. Mendelsohn, L. Rainer, G. Todesco, D. Shipley, and M. Adelaar. 2004. *Emerging energy-saving technologies and practices for the buildings sector as of 2004.* ACEEE report A042. Washington, DC: American Council for an Energy-Efficient Economy.

Shipley, A.M., and R.N. Elliott. 2006. *Ripe for the picking: Have we exhausted the low-hanging fruit in the industrial sector?* ACEEE report IE061. Washington, DC: American Council for an Energy-Efficient Economy.

Sudarshan, A., and J. Sweeney. 2008. *Deconstructing the "Rosenfeld Curve".* Stanford, CA: Stanford University. Online at *http://piee.stanford.edu/cgi-bin/docs/publications/sweeney/Deconstructing%20the%20Rosenfeld%20Curve.pdf.*

Success Story: The Two-fer—How Midwesterners Are Saving Money while Cutting Carbon Emissions

Buntjer, J. 2007. Viking Terrace nominated for national honor. *Worthington Daily Globe,* August 3.

City of Chicago. 2008. *Chicago climate action plan,* 22. Online at *http://www.chicagoclimateaction.org/filebin/pdf/finalreport/CCAPREPORTFINAL.pdf.*

Cuyahoga Community Land Trust. 2008. Energy efficient, accessible, affordable: The green cottages. Cleveland, OH. Online at *http://www.cclandtrust.org/GreenCottages/Efficiency.html,* accessed on October 30, 2008.

Dawson, D. 2008. Personal communication. November 17. Diana Dawson is outreach coordinator for the Cuyahoga County Land Trust, the nonprofit developer of the Cleveland EcoVillage Green Cottages.

Energy Information Administration (EIA). 2008c. *Short term energy and winter fuels outlook.* Table WF01. Washington, DC: U.S. Department of Energy. Online at *http://www.eia.doe.gov/emeu/steo/pub/oct08.pdf,* accessed on October 7, 2008.

Energy Star. 2009. *Reflective roof products.* Washington, DC: U.S. Environmental Protection Agency and U.S. Department of Energy. Online at *http://www.energystar.gov/index.cfm?c=roof_prods.pr_roof_products,* accessed on January 26, 2009.

Hansen, K. 2008. Can Cleveland bring itself back from the brink? *Grist,* May 15. Online at *http://www.grist.org/feature/2008/05/15/cleveland.*

Lobel, H. 2007. Low rent high tech. *Utne Reader,* November/December. Online at *http://www.utne.com/2007-11-01/Low-Rent-High-Tech.aspx.*

Lopez, J. 2008. Personal communication. November 17. Jorge Lopez works for the Southwest Minnesota Housing Partnership and was senior program manager for the Viking Terrace renovation.

Metcalf, M. 2008. Personal communication. November 17. Mandy Metcalf served as the Cleveland EcoVillage project director for the Detroit Shoreway Community Development Organization during the EcoVillage construction.

Minnesota Green Communities. No date. *Viking Terrace Apartments.* Online at *http://www.mngreencommunities.org/projects/profiles/VikingTerrace.pdf,* accessed on October 30, 2008.

Office of Energy Efficiency and Renewable Energy (EERE). 2009a. *Energy savers: Tips on saving energy & money at home: Water heating.* Washington, DC: U.S. Department of Energy. Online at *http://www1.eere.energy.gov/consumer/tips/water_heating.html,* accessed on January 26, 2009.

Success Story: Three Companies Find Efficiency a Profitable Business Strategy

Bechtold, B. 2008a. Triple bottom line effects on people, planet, and profits. Presentation. Online at *http://www.harbec.com/environmental_sustainability.html.*

Bechtold, B. 2008b. Personal communication. November 18. Bob Bechtold is president and CEO of Harbec Plastics.

Corporate Social Responsibility (CSR). 2007. Cogeneration at SC Johnson prevents greenhouse gas emissions equivalent to driving around Earth 7,630 times. Online at *http://www.csrwire.com/News/9492.html*, accessed on April 3, 2009.

DuPont. 2008. 2008 sustainability progress report. Online at *http://www2.dupont.com/Sustainability/en_US/index.html*, accessed on April 3, 2009.

Environmental Protection Agency (EPA). 2009. Climate leaders: Partner profile–SC Johnson. Washington, DC. April 17. Online at *http://www.epa.gov/stateply/partners/partners/scjohnson.html*.

Hoffman, A.J. 2006. *Getting ahead of the curve: Corporate strategies that address climate change.* Washington, DC: Pew Center on Global Climate Change, 90–92. Online at *http://www.pewclimate.org/global-warming-in-depth/all_reports/corporate_strategies*.

Houghton, J.T., G.J. Jenkins, and J.J. Ephraums. 1990. *First assessment report: Scientific assessment of climate change.* Intergovernmental Panel on Climate Change (IPCC). Cambridge, UK: Cambridge University Press.

SC Johnson & Son. 2008. Doing what's right: 2008 public report. Racine, WI, 20–25. Online at *http://www.scjohnson.com/environment/2008_Public_Report.asp*.

Chapter 5
Flipping the Switch to Cleaner Electricity

American Solar Energy Society (ASES). 2007. *Tacking climate change in the U.S.: Potential carbon emissions reductions from energy efficiency and renewable energy by 2030*, edited by C.F. Kutscher. Boulder, CO. Online at *http://www.ases.org/images/stories/file/ASES/climate_change.pdf*.

American Wind Energy Association (AWEA). 2009a. Wind power trends to watch for in 2009. Press release, December 29. Online at *http://www.awea.org/newsroom/releases/trends_to_watch_09_29Dec08.html*.

American Wind Energy Association (AWEA). 2009b. U.S. wind energy projects as of March 31, 2009. Online at *http://www.awea.org/projects*, accessed on May 19, 2009.

Blackwell, D., and M. Richards. 2006. 6 km temperature map. Dallas, TX: SMU Geothermal Laboratory.

California Energy Commission (CEC). 2008. Geothermal energy in California. Sacramento, CA. Online at *http://www.energy.ca.gov/geothermal*, accessed on April 5, 2009.

Cambridge Energy Research Associates (CERA). 2008. *Power capital and costs.* Cambridge, MA.

Creyts, J., A. Derkach, S. Nyquist, K. Ostrowski, and J. Stephenson. 2007. *Reducing U.S. greenhouse gas emissions: How much at what cost?* New York: McKinsey and Company.

Dixon, D., and R. Bedard. 2007. *Assessment of waterpower potential and development needs*, EPRI Report #1014762. Palo Alto, CA: Electric Power Research Institute, Table 3-1.

Electric Power Research Institute (EPRI). 2008. Role of renewable energy in a sustainable generation portfolio (EPRI-NESSIE). Presentation by Thomas Key at the National Wind Coordinating Collaborative (NWCC) Environment Cost & Benefits Workshop, October 8, Washington, DC.

Electric Power Research Institute (EPRI). 2007. *The power to reduce CO_2 emissions: The full portfolio.* Discussion paper prepared for the EPRI 2007 Summer Seminar attendees. Palo Alto, CA.

Energy Information Administration (EIA). 2009a. *Electric power annual.* Washington, DC: U.S. Department of Energy. Online at *http://www.eia.doe.gov/cneaf/electricity/epa/epa_sum.html*.

Energy Information Administration (EIA). 2009b. Impacts of a 25-percent renewable electricity standard as proposed in the American Clean Energy and Security Act discussion draft. Washington, DC: U.S. Department of Energy.

Energy Information Administration (EIA). 2008. Energy market and economic impacts of S.2191, the Lieberman-Warner Climate Security Act of 2007. Washington, DC: U.S. Department of Energy.

Energy Information Administration (EIA). 2008d. *Emissions of greenhouse gases in the United States 2007.* DOE/EIA-0573(2007). Washington, DC: U.S. Department of Energy. Online at *ftp://ftp.eia.doe.gov/pub/oiaf/1605/cdrom/pdf/ggrpt/057307.pdf*.

Energy Information Administration (EIA). 2008e. Renewable energy consumption and electricity preliminary 2007 statistics. Table 3. Washington, DC: U.S. Department of Energy. Online at *http://www.eia.doe.gov/cneaf/alternate/page/renew_energy_consump/table3.html*.

Energy Information Administration (EIA). 2007. *Energy and economic impacts of implementing both a 25-percent renewable portfolio standard and a 25-percent renewable fuel standard by 2025.* Washington, DC: U.S. Department of Energy.

Energy Information Administration (EIA). 1986. *Analysis of nuclear power plant construction costs.* DOE/EIA-0485. Washington, DC: U.S. Department of Energy.

Environmental Protection Agency (EPA). 2008a. *EPA analysis of the Lieberman-Warner climate security act of 2008.* Washington, DC. Online at *http://www.epa.gov/climatechange/downloads/s2191_EPA_Analysis.pdf*.

Environmental Protection Agency (EPA). 2008d. *Landfill gas energy projects and candidate landfills.* Washington, DC. Online at *http://epa.gov/lmop/docs/map.pdf*.

Erickson, W., G. Johnson, M.D. Strickland, D. Young Jr., K. Sernka, and R. Good. 2001. *Avian collisions with wind turbines: A summary of existing studies and comparisons of avian collision mortality in the United States.* Prepared by Western Ecosystems Technology, Inc. Washington, DC: National Wind Coordinating Committee. Online at *http://www. nationalwind.org/publications/wildlife/avian_collisions.pdf.*

Freese, B., S. Clemmer, and A. Nogee. 2008. *Coal power in a warming world: A sensible transition to cleaner energy options.* Cambridge, MA: Union of Concerned Scientists. Online at *http://www.ucsusa.org/assets/documents/clean_energy/Coal-power-in-a-warming-world.pdf.*

Garwin, R.L. 2001. *Can the world live without nuclear power?* Presentation at the Nuclear Control Institute, April 9. Washington, DC.

Geothermal Energy Association (GEA). 2008a. Washington, DC. Online at *http://www.geo-energy.org/aboutGE/currentUse. asp#world.*

Geothermal Energy Association (GEA). 2008b. U.S. geothermal power and production and development update: August 2008. Washington, DC. Online at *http://www.geo-energy.org/publications/reports/Geothermal_Update_August_7_2008_FINAL.pdf.*

Geothermal Energy Association (GEA). 2008c. State geothermal resource maps. Washington, DC. Online at *http://www.geo-energy.org/information/resources.asp.*

Global Wind Energy Council (GWEC). 2008. *Global wind 2008 report.* Brussels.

Goldberg, M. 2000. *Federal energy subsidies: Not all technologies are created equal.* Washington, DC: Renewable Energy Policy Project.

Google. 2008. Unleashing the heat beneath our feet: How the U.S. government can support enhanced geothermal systems (EGS). Online at *http://www.google.org/egs,* accessed on April 13, 2009.

Government Accountability Office (GAO). 2006. *Department of Energy: Key challenges remain for developing and deploying advanced energy technologies to meet future needs.* GAO-07-106. Washington, DC. Online at: *http://www.gao.gov/new.items/d07106.pdf,* accessed on January 28, 2009.

The Guardian. 2009. Areva clashes with Finnish utility over delays to new nuclear plant. January 14.

Holttinen, H., P. Meibom, C. Ensslin, L. Hofmann, A. Tuohy, J.O. Tande, A. Astanqueiro, E. Gomez, L. Soder, A. Shakoor, J.C. Smith, B. Parsons, and F. van Hulle. 2007. *State of the art of design and operation of power systems with large amounts of wind power: Summary of IEA wind collaboration.* EWEC 2007 conference, May 7–10, Milan. Online at *http://www.risoe.dk/rispubl/art/2007_120_paper.pdf.*

Intergovernmental Panel on Climate Change (IPCC). 2005. *IPCC special report on carbon dioxide capture and storage.* Prepared by Working Group III of the Intergovernmental Panel on Climate Change, edited by B. Metz, B., O. Davison, H.C. de Coninck, M. Loos, and L.A. Meyer. Cambridge, UK: Cambridge University Press.

International Atomic Energy Agency (IAEA). 2008. *Uranium report: Plenty more where that came from.* Staff report. Vienna. Online at *http://www.iaea.org/NewsCenter/News/2008/uraniumreport.html.*

International Energy Agency (IEA). 2007. *Renewables for heating and cooling: Untapped potential.* Paris: Renewable Energy Technology Deployment. Online at *http://www.iea-retd.org/files/Heating_Cooling_Final_WEB.pdf.*

The Keystone Center. 2007. *Nuclear power joint fact-finding.* Keystone, CO. Online at *http://www.ne.doe.gov/pdfFiles/rpt_KeystoneReportNuclearPowerJointFactFinding_2007.pdf.*

Kutscher, C. 2008. Tackling climate change: Can we afford it? *Solar Today,* March/April.

Massachusetts Institute of Technology (MIT). 2007. *The future of coal: Options for a carbon-constrained world.* Cambridge, MA.

Massachusetts Institute of Technology (MIT). 2003. The future of nuclear power. Cambridge, MA. Online at *http://web.mit.edu/nuclearpower.*

Milbrandt, A. 2005. *A geographic perspective on the current biomass resource availability in the United States.* Golden, CO: U.S. Department of Energy, Office of Energy Efficiency and Renewable Energy. Online at *http://www.nrel.gov/docs/fy06osti/39181.pdf.*

Miller, C.C. 2008. Thin film solar companies raise hundreds of millions in financing. *New York Times,* September 11.

National Energy Technology Laboratory (NETL). 2007. *Cost and performance baseline for fossil energy plants, volume 1: Bituminous coal and natural gas to electricity.* Final report, revision 1. Pittsburgh, PA: U.S. Department of Energy.

National Energy Technology Laboratory (NETL). 2006. *Carbon sequestration atlas of the United States and Canada.* Pittsburgh, PA: U.S. Department of Energy.

National Renewable Energy Laboratory (NREL). 2007. Concentrating solar power prospects in the southwest United States. Golden, CO. Online at *http://www.nrel.gov/csp/images/3pct_csp_sw.jpg.*

National Renewable Energy Laboratory (NREL). 2005. *Biomass maps: Biomass resources available in the United States.* Golden, CO: U.S. Department of Energy. Online at *http://www.nrel.gov/gis/biomass.html.*

Navigant Consulting. 2008. *Economic impacts of extending federal solar tax credits.* Final report prepared for the Solar Energy Research and Education Foundation (SEREF). September.

Nogee, A., J. Deyette, and S. Clemmer. 2007. The projected impacts of a national renewable portfolio standard. *Electricity Journal* 20(4).

Office of Energy Efficiency and Renewable Energy (EERE). 2008. *20% wind energy by 2030: Increasing wind energy's contribution to U.S. electricity supply.* DOE/GO-102008-2578. Washington, DC: U.S. Department of Energy. Online at *http://www.20percentwind.org/20p.aspx?page=Report.*

Office of Energy Efficiency and Renewable Energy (EERE). 2008a. *An evaluation of enhanced geothermal systems technology.* Washington, DC: U.S. Department of Energy. Online at *http://www1.eere.energy.gov/geothermal/pdfs/evaluation_egs_tech_2008.pdf.*

Organization for Economic Cooperation and Development (OECD). 2008. *Nuclear energy outlook.* OECD Publishing. Online at *http://www.oecdbookshop.org/oecd/display.asp?K=5KZ G1MRNZGJH&lang=EN&sort=sort_date%2Fd&sf1=Title&st 1=nuclear+energy+outlook&sf3=SubjectCode&st3=41&st4=not +E4+or+E5+or+P5&sf4=SubVersionCode&ds=nuclear+energy +outlook%3B+Nuclear+Energy%3B+&m=1&dc=2&plang=en.*

Perlack, R.D., L.L. Wright, A.F. Turhollow, R.L. Graham, B.J. Stokes, and D.C. Erbach. 2005. *Biomass a feedstock for a bioenergy and bioproducts industry: The technical feasibility of a billion-ton annual supply.* DOE/GO-102005-2135. Oak Ridge, TN: Oak Ridge National Laboratory.

Peters, R., and L. O'Malley. 2008. Storing renewable power. In *Making renewable energy a priority.* Calgary: The Pembina Institute.

Schlissel, D., M. Mullet, and R. Alvarez. 2009. *Nuclear loan guarantees: Another taxpayer bailout ahead?* Cambridge, MA: Union of Concerned Scientists.

Sissine, F. 1994. *Renewable energy: A national commitment?* Washington, DC: Congressional Research Service, Science Policy Research Division.

Solar Energy Technologies Program (SETP). 2007. *Concentrating solar power: FY09 proposed solar initiative.* Budget summit meeting, National Press Club, March. Washington, DC: U.S. Department of Energy.

Solarbuzz. 2008. *Marketbuzz 2008: Annual world solar photovoltaic industry report.* Online at *http://www.solarbuzz.com/Marketbuzz2008-intro.htm*, accessed on February 10, 2009.

Tester, J.W., S. Petty, J. Garnish, A. Batchelor, L. Drake, and R. Veatch. 2006. *The future of geothermal energy: Energy recovery from enhanced/engineered geothermal systems (EGS), assessment of impact for the U.S. by 2050.* Final report to the United States Department of Energy Geothermal Program. Cambridge, MA: Massachusetts Institute of Technology.

Union of Concerned Scientists (UCS). 2009. Clean power, green jobs. Cambridge, MA.

Union of Concerned Scientists (UCS). 2007. Experts agree: Renewable electricity standards are a key driver of new renewable energy development. Cambridge, MA. Online at *http://www.ucsusa.org/clean_energy/solutions/renewable_energy_solutions/experts-agree-renewable.html.*

Union of Concerned Scientists (UCS). 2007a. *Nuclear power in a warming world.* Cambridge, MA.

Union of Concerned Scientists (UCS). 2003. How nuclear power works. Cambridge, MA. Online at *http://www.ucsusa.org/nuclear_power/nuclear_power_technology/how-nuclear-power-works.html.*

Williams, C.F., M.J. Reed, R.H. Mariner, J. DeAngelo, and S.P. Galanis Jr. 2008. *Assessment of moderate- and high-temperature geothermal resources of the United States.* U.S. Geological Survey fact sheet 2008-3082, 4. Washington, DC: U.S. Department of the Interior.

Wiser, R., and M. Bolinger. 2008. *Annual report on U.S. wind power installation, cost, and performance trends 2007.* Berkeley, CA: Lawrence Berkeley National Laboratory. Online at *http://eetd.lbl.gov/EA/emp/reports/lbnl-275e.pdf.*

Success Story: The Little Country that Could

Danish Energy Agency. 2008. *Energy statistics 2007.* Copenhagen. Online at *http://www.energistyrelsen.dk/graphics/UK_Facts_Figures/Statistics/yearly_statistics/2007/energy%20statistics %202007%20uk.pdf*, accessed on January 28, 2009.

Danish Wind Energy Association (DWEA). 2009. Turbines in Denmark. Online at *http://www.windpower.org/composite-1458.html*, accessed on January 28, 2009.

Freese, B., S. Clemmer, and A. Nogee. 2008. *Coal power in a warming world: A sensible transition to cleaner energy options.* Cambridge, MA: Union of Concerned Scientists. Online at *http://www.ucsusa.org/assets/documents/clean_energy/Coal-power-in-a-warming-world.pdf.*

Government Accountability Office (GAO). 2006. *Department of Energy: Key challenges remain for developing and deploying advanced energy technologies to meet future needs.* GAO-07-106. Washington, DC. Online at *http://www.gao.gov/new.items/d07106.pdf*, accessed on January 28, 2009.

Ministry of Climate and Energy. 2008. The Danish example. Copenhagen. Online at: *http://www.kemin.dk/en-US/facts/danishexample/Documents/The%20Danish%20Case.pdf*, accessed on January 28, 2009.

Success Story: Surprises in the Desert

ACCIONA. 2009. Nevada Solar One. Online at *http://www.acciona-na.com/About-Us/Our-Projects/U-S-/Nevada-Solar-One.aspx*, accessed on April 29, 2009.

California Energy Commission (CEC). 2008a. Large solar energy projects: Solar thermal projects under review or announced. Sacramento, CA. Online at *http://www.energy.ca.gov/siting/solar/index.html*, accessed on August 25, 2008.

Energy Information Administration (EIA). 2008f. *Solar thermal collector manufacturing activities 2007.* Washington, DC: U.S. Department of Energy. Online at *http://www.eia.doe.gov/cneaf/solar.renewables/page/solarreport/solar.html.*

Stoddard, L., J. Abiecunas, and R. O'Connell. 2006. *Economic, energy, and environmental benefits of concentrating solar power in California.* NREL/SR-550-39291. Golden, CO: National Renewable Energy Laboratory. Online at *http://www.nrel.gov/csp/pdfs/39291.pdf.*

Chapter 6
You Can Get There From Here: Transportation

American Association of State and Highway Transportation Officials (AASHTO). 2007. *Transportation: Invest in our future.* Washington, DC. Online at *http://www.transportation1.org/tif5report/TIF5.pdf.*

American Association of State and Highway Transportation Officials (AASHTO). 2003. *Transportation: Invest in our future.* Washington, DC. Online at *http://freight.transportation.org/doc/FreightRailReport.pdf.*

American Honda Motor Company. 2009. Specifications. Online at *http://automobiles.honda.com/fcx-clarity/specifications.aspx,* accessed on February 20, 2009.

American Petroleum Institute (API). 2000. *Technology vision 2020: A report on technology and the future of the U.S. petroleum industry.* Washington, DC. Online at *http://www1.eere.energy.gov/industry/petroleum_refining/pdfs/techvision.pdf.*

American Solar Energy Society (ASES). 2007. *Tackling climate change in the U.S.: Potential carbon emissions reductions from energy efficiency and renewable energy by 2030,* edited by C.F. Kutscher. Boulder, CO. Online at *http://www.ases.org/images/stories/file/ASES/climate_change.pdf.*

An, F., F. Stodolsky, A. Vyas, R. Cuenca, and J.J. Eberhardt. 2000. Scenario analysis of hybrid class 3-7 heavy vehicles. Document number 2000-01-0989. Presented at SAE 2000 World Congress. Warrendale, PA: Society of Automotive Engineers. Online at *http://www.sae.org/technical/papers/2000-01-0989*

Anair, D. 2008. *Delivering the green: Reducing trucks' climate impacts while saving at the pump.* Cambridge, MA: Union of Concerned Scientists.

Anden, A., M. Ruth, K. Isben, J. Jechura, K. Neeves, J. Sheehan, and B. Wallace. 2002. *Lignocellulosic biomass to ethanol process design and economics utilizing co-current dilute acid prehydrolysis and enzymatic hydrolysis for corn stover.* Golden, CO: National Renewable Energy Laboratory. Online at *http://www.nrel.gov/docs/fy02osti/32438.pdf.*

Argonne National Laboratory (ANL). 2008. The greenhouse gases, regulated emissions, and energy use in transportation (GREET) model. Version 1.8b. Argonne, IL. Online at *http://www.transportation.anl.gov/publications/transforum/v8/v8n2/greet_18b.html.*

Bandivadekar, A., K. Bodek, L. Cheah, C. Evans, T. Groode, J. Heywood, E. Kasseris, M. Kromer, and M. Weiss. 2008. *On the road in 2035: Reducing transportation's petroleum consumption and GHG emissions.* LFEE 2008-05 RP. Cambridge, MA: Massachusetts Institute of Technology Laboratory for Energy and the Environment.

Bartis, J.T., F. Camm, and D.S. Ortiz. 2008. *Producing liquid fuels from coal: Prospects and policy issues.* Santa Monica, CA: RAND Corporation.

Bedsworth, L.W. 2004. *Climate control: Global warming solutions for California cars.* Cambridge, MA: Union of Concerned Scientists. Online at *http://www.ucsusa.org/assets/documents/clean_vehicles/climate_control_-lo-res.pdf.*

Bemis, G. 2008. Developing a methodology to allocate AB 118 program funds for LDVs. Presentation at staff workshop, September 19. Sacramento, CA: California Energy Commission. Online at *http://www.energy.ca.gov/proceedings/2008-ALT-1/documents/2008-09-19_workshop/presentations/Gerrys_Bemis.pdf.*

Bordoff, J.E., and P.J. Noel. 2008. Pay-as-you-drive auto insurance: A simple way to reduce driving-related harms and increase equity. Washington, DC: The Brookings Institution, July. Online at *http://www.brookings.edu/papers/2008/07_payd_bordoffnoel.aspx,* accessed on February 20, 2009.

Brons, M., P. Nijkamp, E. Pels, and P. Rietveld. 2006. *A meta-analysis of the price elasticity of gasoline demand: A system of equations approach.* TI 2006-106/3. Amsterdam: Tinbergen Institute. Online at *http://ftp.tinbergen.nl/discussionpapers/06106.pdf.*

California Air Resources Board (CARB). 2009. *Proposed regulation to implement the low carbon fuel standard.* Volume I staff report: Initial statement of reasons. Sacramento, CA. Online at *http://www.arb.ca.gov/fuels/lcfs/030409lcfs_isor_vol1.pdf.*

California Air Resources Board (CARB). 2008. *Comparison of greenhouse gas reductions for the United States and Canada under U.S. CAFE standards and California Air Resources Board greenhouse gas regulations: An enhanced technical assessment.* Online at *http://www.climatechange.ca.gov/publications/arb/ARB-1000-2008-012/ARB-1000-2008-012.PDF.*

California Air Resources Board (CARB). 2008a. *Climate change proposed scoping plan appendices.* Volume I supporting documents and detail. Sacramento, CA. Online at *http://www.arb.ca.gov/cc/scopingplan/document/appendix1.pdf.*

Census Bureau. 2002. Vehicle inventory and use survey: Microdata. Online at *http://www.census.gov/svsd/www/vius/2002.html,* accessed on April 3, 2009.

Center for Transportation Analysis (CTA). 2008. *Transportation energy data book*, edition 27. Oak Ridge, TN: Oak Ridge National Laboratory. Online at *http://cta.ornl.gov/data/index.shtml*.

Cooper, C., R. Reinhart, M. Kromer, R. Wilson, and R. Schubert. Forthcoming. *Reducing heavy-duty vehicle fuel consumption and GHG emissions*, edited by P. Miller. Boston, MA: Northeast States Center for a Clean Air Future (NESCCAF); San Francisco, CA: International Council on Clean Transportation (ICCT).

Cowart, W. 2008. How much can we slow VMT growth? Presented to CCAP VMT & Climate Change Workshop. Cambridge, MA: Cambridge Systematics. Online at *http://www.ccap.org/docs/resources/460/How%20much%20can%20we%20slow%20VMT%20growth%20May%202008.pdf*.

Dahl, C. 1993. *A survey of energy demand elasticities in support of the development of the NEMS*. De-AP01-93EI23499. U.S. Department of Energy.

Department of Energy (DOE). 2000. *Technology roadmap for the 21st century truck program*. 21CT-001. Washington, DC.

Department of Transportation (DOT). 2000. Comprehensive truck size and weight study. Final report. Volume 2, Chapter 3, Table III-4. Washington, DC. Online at *http://www.fhwa.dot.gov/policy/otps/truck/finalreport.htm*.

Department of Transportation (DOT), Department of Energy (DOE), and Environmental Protection Agency (EPA). 2002. *Effects of the alternative motor fuels act: CAFE incentives policy*. Report to Congress. Washington, DC. Online at *http://www.nhtsa.gov/cars/rules/rulings/CAFE/alternativefuels/index.htm*.

Duvall, M. 2002. *Comparing the benefits and impacts of hybrid electric vehicle options for compact sedan and sport utility vehicles*. 1006892. Palo Alto, CA: Electric Power Research Institute. Online at *http://mydocs.epri.com/docs/public/000000000001006892.pdf*.

Electric Power Research Institute (EPRI). 2007a. *Environmental assessment of plug-in hybrid electric vehicles*. Volume 1: Nationwide greenhouse gas emissions. 1015325. Palo Alto, CA.

Energy Information Administration (EIA). 2009. *Annual energy outlook 2009 with projections to 2030*. Washington, DC: U.S. Department of Energy. Online at *http://www.eia.doe.gov/oiaf/aeo/aeoref_tab.html*.

Energy Information Administration (EIA). 2009c. Petroleum navigator: Annual U.S. regular all formulations retail gasoline prices. Washington, DC: U.S. Department of Energy. Online at *http://tonto.eia.doe.gov/dnav/pet/hist/mg_rt_usa.htm*.

Energy Information Administration (EIA). 2009d. Petroleum navigator: Annual U.S. all grades all formulations retail gasoline prices. Washington, DC: U.S. Department of Energy. Online at *http://tonto.eia.doe.gov/dnav/pet/hist/mg_tt_usa.htm*.

Energy Information Administration (EIA). 2009e. *Short term energy outlook*. Table S7. Washington, DC: U.S. Department of Energy. Online at *http://www.eia.doe.gov/emeu/states/sep_sum/html/sum_btu_tra.html*, accessed on April 8, 2009.

Energy Information Administration (EIA). 2008a. *Annual energy outlook 2008 with projection to 2030*. Washington, DC: U.S. Department of Energy.

Energy Information Administration (EIA). 2007. *Energy and economic impacts of implementing both a 25-percent renewable portfolio standard and a 25-percent renewable fuel standard by 2025*. Washington, DC: U.S. Department of Energy.

Environmental Protection Agency (EPA). 2008. Inventory of U.S. greenhouse gas emissions and sinks: 1990–2006. EPA 430-R-08-005. Washington, DC. Online at *http://www.epa.gov/climatechange/emissions/usinventoryreport.html*.

Espey, M. 2004. Gasoline demand revisited: An international meta-analysis of elasticities. *Transport Reviews* 24(3):275–292.

Ewing, R., K. Bartholomew, S. Winkelman, J. Walters, and D. Chen. 2007. Growing cooler: The evidence on urban development and climate change. Washington, DC: Urban Land Institute. Online at *http://www.smartgrowthamerica.org/gcindex.html*.

Ewing, R., R. Pendall, and D. Chen. 2002. *Measuring sprawl and its impact*. Washington, DC: Smart Growth America. Online at *http://www.smartgrowthamerica.org/sprawlindex/MeasuringSprawl.PDF*.

Fargione, J., J. Hill, D. Tillman, S. Polasky, and P. Hawthorne. 2008. Land clearing and the biofuel carbon debt. *Science* 319(5867).

Ford Motor Company. 2009. Ford, EPRI add 7 new utility partners, battery maker to plug-in hybrid vehicle program. Press release, February 3. Online at *http://www.ford.com/about-ford/news-announcements/press-releases/press-releases-detail/pr-ford-epri-add-7-new-utility-29804*.

Friedman, D., C.E. Nash, and C. Ditlow. 2003. *Building a better SUV: A blueprint for saving lives, money, and gasoline*. Cambridge, MA: Union of Concerned Scientists.

General Motors (GM). 2008. Saturn Vue Green Line plug-in hybrid SUV may begin production in 2010. Press release, January 14.

Gonzales, A. 2008. General Motors recharges on future of electric car. *Seattle Times*, June 18. Online at *http://seattletimes.nwsource.com/html/businesstechnology/2008003356_electriccar18.html*, accessed on February 19, 2009.

Goodwin, P., J. Dargay, and M. Hanly. 2004. *Elasticities of road traffic and fuel consumption with respect to price and income: A review*. 24(3):275–292. London: University College London.

Gordon, D., D.L. Greene, M.H. Ross, and T.P. Wenzel. 2007. *Sipping fuel and saving lives: Increasing fuel economy without sacrificing safety.* Prepared for The International Council on Clean Transportation.

Greene, D., J. German, and M. Delucchi. 2009. Automotive fuel economy: The case for market failure. Chapter 11 in *Reducing climate impacts in the transportation sector*, edited by D. Sperling and J.S. Cannon. Dordrecht, Netherlands: Springer.

Greene, N., F.E. Celik, B. Dale, M. Jackson, K. Jayaward-hana, H. Jin, E.D. Larson, M. Laser, L. Lynd, D. MacKensie, J. Mark, J. McBride, S. McLaughlin, and D. Saccardi. 2004. *Growing energy: How biofuels can help end America's oil dependence.* New York: Natural Resources Defense Council. Online at *http://www.nrdc.org/air/energy/biofuels/biofuels.pdf.*

Hill, W.R. 2003. HFC152a as the alternative refrigerant. Presentation at the MACSummit, Brussels, February 11. Online at *http://europa.eu.int/comm/environment/air/mac2003/pdf/hill.pdf.*

Interlaboratory Working Group (IWG). 2000. *Scenarios for a clean energy future.* ORNL/CON-476 and LBNL-44029. Oak Ridge, TN: Oak Ridge National Laboratory; and Berkeley, CA: Lawrence Berkeley National Laboratory.

Kalhammer, F.R., B.M. Kopf, D.H. Swan, V.P. Roan, and M.P. Walsh. 2007. Status and prospects for zero emissions vehicle technology: Report of the ARB independent expert panel 2007. Prepared for the State of California Air Resources Board. Online at *http://www.arb.ca.gov/msprog/zevprog/zevreview/zev_panel_report.pdf.*

Kliesch, J. 2008. *Setting the standard: How cost-effective technology can increase vehicle fuel economy.* Cambridge, MA: Union of Concerned Scientists.

Komatsu, M., T. Takaoka, T. Ishikawa, N. Suzuki, Y. Gotoda, and T. Ozawa. 2008. *Study on the potential benefits of a plug-in hybrid system.* SAE 2008-01-0456. Warrendale, PA: Society of Automotive Engineers.

Lowell, D., and T. Balon. 2009. *Setting the stage for regulation of heavy-duty fuel economy & GHG emissions: Issues and opportunities.* Prepared by MJ Bradley and Associates LLC. Washington, DC: International Council on Clean Transportation.

Marland, G., T.A. Boden, and R.J. Andres. 2008. Global, regional, and national fossil fuel CO_2 emissions. In *Trends: A compendium of data on global change.* Oak Ridge, TN: Oak Ridge National Laboratory Carbon Dioxide Information Analysis Center.

Mathews, H. 2008. J.B. Hunt's sustainability efforts. Presented at the Improving the Fuel Economy of Heavy-Duty Fleets II Conference. San Diego.

McManus, W. 2007. *Economic analysis of feebates to reduce greenhouse gas emissions from light vehicles for California.* Ann Arbor, MI: University of Michigan Transportation Research Institute. Online at *http://www.umtri.umich.edu/content/UMTRI-2007-19-2.pdf.*

Morrow, K., D. Karner, and J. Francfort. 2008. *Plug-in hybrid electric vehicle charging infrastructure review.* INL/EXT-08-15058. Idaho Falls, ID: Idaho National Laboratory, Office of Energy Efficiency and Renewable Energy, U.S. Department of Energy.

National Center for Transit Research (NCTR). 2007. *Public transit in America: Analysis of access using the 2001 national household travel survey.* BD 549-30. Tampa, FL: University of South Florida Center for Urban Transportation Research.

National Research Council (NRC). 2008. *Transitions to alternative transportation technologies: A focus on hydrogen.* Washington, DC: The National Academies Press.

National Research Council (NRC). 2002. *Effectiveness and impact of Corporate Average Fuel Economy (CAFE) standards.* The National Academies Press.

Office of Energy Efficiency and Renewable Energy (EERE). 2009c. Alternative fueling station total counts by state and fuel type. Washington, DC: U.S. Department of Energy. Online at *http://www.afdc.energy.gov/afdc/fuels/stations_counts.html,* accessed on April 8, 2009.

Office of Transportation and Air Quality (OTAQ). 2008. *Light-duty automotive technology and fuel economy trends: 1975–2008.* Ann Arbor, MI: U.S. Environmental Protection Agency. Online at *http://www.epa.gov/OMS/cert/mpg/fetrends/420r08015.pdf.*

Perlack, R., L.L. Wright, A.F. Turhollow, R.L. Graham, B.J. Stokes, and D.C. Erbach. 2005. *Biomass as feedstock for a bioenergy and bioproducts industry: The technical feasibility of a billion-ton annual supply.* TM-2005-66. Oak Ridge, TN: Oak Ridge National Laboratory.

Pettersen, J. 2003. Pros and cons of the options: Improved 134a, HFC-152a, and CO_2 (R-744). Presentation at the MACSummit, Brussels, February 11. Online at *http://europa.eu.int/comm/environment/air/mac2003/pdf/pettersen.pdf.*

Santini, D., and A.Vyas. 2008. *More complications in estimation of oil savings via electrification of light duty vehicles.* Presented at the PLUG-IN 2008 Conference, McEnery Convention Center, San Jose, CA, July 21–24. Argonne National Laboratory. Online at *http://www.transportation.anl.gov/pdfs/HV/524.pdf.*

Searchinger, T., R. Heimlich, R.R. Houghton, F. Dong, A. Elobeid, J. Fabiosa, S. Tokgoz, D. Hayes, and T. Yu. 2008. Use of U.S. croplands for biofuels increases global warming pollution from land-use change. *Science* 319(5867):1238–1240.

Small, K., and K. Van Dender. 2006. *Fuel efficiency and motor vehicle travel: The declining rebound effect*. Working paper 050603. Irvine, CA: University of California–Irvine Department of Economics.

Society of Automotive Engineers (SAE). 2008. Industry evaluation of low global warming potential refrigerant HFO1234yf. Warrendale, PA. Online at *http://www.sae.org/standardsdev/tsb/cooperative/crp1234summary.pdf*.

Sperling, D., and D. Gordon. 2009. *Two billion cars: Driving toward sustainability*. New York: Oxford University Press.

Stodolsky, F. 2002. *Railroad and locomotive technology roadmap*. Argonne, IL: Argonne National Laboratory Center fo. Transportation Research.

Sun, L.H. 2008. New ridership record shows U.S. still lured to mass transit. *Washington Post*, December 8.

Task Force on Strategic Unconventional Fuels. 2006. *Development of America's strategic unconventional fuel resources: Initial report to the president and the Congress of the United States*. Online at *http://www.fossil.energy.gov/programs/reserves/npr/publications/sec369h_report_epact.pdf*.

Tate, E.D., M. Harpster, and P.J. Savagian. 2008. *The electrification of the automobile: From conventional hybrid, to plug-in hybrids, to extended-range electric vehicles*. SAE 2008-01-0458. Warrendale, PA: Society of Automotive Engineers.

Toyota Motor Sales. 2009. Toyota maintains pace, broadens scope of advanced environmental technologies. Press release, January 10. Online at *http://pressroom.toyota.com/pr/tms/toyota/maintain-pace-broaden-scope.aspx*, accessed on April 3, 2009.

Turrentine, T., and K. Kurani. 2000. *Progress in electric vehicle technology and electric vehicles from 1990 to 2000: The role of California's zero emission vehicle production requirement*. UCD-ITS-RR-00-20. Davis, CA: University of California–Davis Institute of Transportation Studies.

Union of Concerned Scientists (UCS). 2008a. Land use changes and biofuels. Cambridge, MA. Online at *http://www.ucsusa.org/assets/documents/clean_vehicles/Indirect-Land-Use-Factsheet.pdf*.

Vyas, A., C. Saricks, and F. Stodolsky. 2003. *The potential effect of future energy-efficiency and emissions-improving technologies on fuel consumption of heavy trucks*. ANL/ESD/02-4. Argonne, IL: Argonne National Laboratory.

Wang, M. 2008. *Estimation of energy efficiencies of U.S. petroleum refineries*. Argonne, IL: Argonne National Laboratory.

Ward's Auto Data. No date. Online at *www.wardsauto.com* (by subscription only).

Worrell, E., and C. Galitsky. 2005. *Energy efficiency improvement and cost saving opportunities for petroleum refineries: An ENERGY STAR® guide for energy and plant managers*. Berkeley, CA: Lawrence Berkeley National Laboratory. Online at *http://ies.lbl.gov/iespubs/energystar/petroleumrefineries.pdf*.

Success Story: Jump-Starting Tomorrow's Biofuels

Bluefire Ethanol. 2008. Bluefire issues annual letter to the shareholders. Online at *http://bluefireethanol.com/pr/65*, accessed on January 30, 2009.

Energy Information Administration (EIA). 2008a. *Annual energy outlook 2008 with projection to 2030*. Washington, DC: U.S. Department of Energy. Online at *http://www.eia.doe.gov/oiaf/archive/aeo08/index.html*.

Energy Information Administration (EIA). 2008g. Number and capacity of operable petroleum refineries by PAD district and state as of January 1, 2008. In *Refinery capacity report*, form EIA-820. Washington, DC: U.S. Department of Energy. Online at *http://www.eia.doe.gov/pub/oil_gas/petroleum/data_publications/refinery_capacity_data/current/refcap08.pdf*.

LaMonica, M. 2008. Inside Mascoma's ethanol-making bug lab. CNET News, September 22. Online at *http://news.cnet.com/8301-11128_3-10047209-54.html*, accessed on January 30, 2009.

Range Fuels. 2007. Our first commercial plant. Online at *http://www.rangefuels.com/our-first-commercial-plant*, accessed on February 3, 2009.

Reidy, C. 2008. Mascoma gets government funding for fuel plant. *Boston Globe*, October 7. Online at *http://www.boston.com/business/ticker/2008/10/mascoma_gets_go.html*, accessed on January 30, 2009.

Renewable Fuels Association (RFA). 2009. *2009 ethanol industry outlook*. Washington, DC. Online at *http://www.ethanolrfa.org/objects/pdf/outlook/RFA_Outlook_2009.pdf*.

Union of Concerned Scientists (UCS). 2008a. Land use changes and biofuels. Cambridge, MA. Online at *http://www.ucsusa.org/assets/documents/clean_vehicles/Indirect-Land-Use-Factsheet.pdf*.

Union of Concerned Scientists (UCS). 2008b. Food as fuel. Cambridge, MA. Online at *http://www.ucsusa.org/assets/documents/clean_vehicles/Food-for-Fuels-Factsheet.pdf*.

Verenium. 2009. Verenium announced first commercial cellulosic ethanol project. January 15. Online at *http://phx.corporate-ir.net/phoenix.zhtml?c=81345&p=RssLanding&cat=news&id=1244987*, accessed on January 30, 2009.

Success Story: It Takes an Urban Village to Reduce Carbon Emissions

American Planning Association (APA). 2007. Great places in America: Streets. Clarendon-Wilson Corridor, Arlington, Virginia. Chicago, IL. Online at *http://www.planning.org/greatplaces/streets/2008/clarendonwilson.htm*, accessed on February 5, 2009.

Arlington Rapid Transit (ART). 2009. About ART-Arlington transit. Arlington County, VA. Online at *http://www.commuterpage.com/art/aboutART.htm*.

Department of Community Planning, Housing, and Development (CPHD). 2008a. Special planning areas: Arlington, VA. Online at *http://www.co.arlington.va.us/ Departments/CPHD/planning/docs/CPHDPlanningDocsGLUP_ metrocorridors.aspx*, accessed on February 5, 2009.

Department of Community Planning, Housing, and Development (CPHD). 2008b. Smart growth: Arlington, VA. Online at *http://www.arlingtonva.us/Departments/CPHD/ planning/CPHDPlanningSmartGrowth.aspx*, accessed on February 5, 2009.

Environmental Protection Agency (EPA). 2005. EPA congratulates Atlanta on smart growth success. Washington, DC. Online at *http://yosemite.epa.gov/opa/admpress.nsf/ 9f9e145a6a71391a852572a000657b5e/0e30c482fa56b3ac 852570d00057768b!OpenDocument*, accessed on February 5, 2009.

Environmental Protection Agency (EPA). 2002. Arlington County, Virginia: National award for smart growth achievement 2002 winners presentation. Washington, DC. Online at *http://www.epa.gov/dced/arlington.htm*, accessed on February 5, 2009.

Reuters. 2007. Bostonians walk to work, Portlanders cycle, but most drive. June 13. Online at *http://www.reuters.com/ article/lifestyleMolt/idUSN1337666520070613*, accessed on February 5, 2009.

Suarez, R. 2008. Population growth burdens roads, schools, and state programs. PBS: The Online NewsHour, October 22. Online at *http://www.pbs.org/newshour/bb/business/july-dec08/blueprint_10-22.html*, accessed on February 5, 2009.

Chapter 7
We Can Do It: Analyzing Climate Solutions

Apollo Alliance. 2008. *The new Apollo program: Clean energy, good jobs: An economic strategy for American prosperity.* San Francisco. Online at *http://apolloalliance.org/apollo-14/the-full-report*.

Birdsey, R.A. 1996. Carbon storage for major forest types and regions in the conterminous United States. In *Forests and global change: forest management opportunities for mitigating carbon emissions*, edited by R.N. Sampson and D. Hair, 379. Washington, DC: American Forests.

Bordoff, J.E., and P.J. Noel. 2008. Pay-as-you-drive auto insurance: A simple way to reduce driving-related harms and increase equity. Washington, DC: The Brookings Institution, July. Online at *http://www.brookings.edu/papers/2008/07_ payd_bordoffnoel.aspx*, accessed on February 20, 2009.

Boucher, D. 2008. *Out of the woods: A realistic role for tropical forests in curbing global warming.* Cambridge, MA: Union of Concerned Scientists.

Congressional Budget Office (CBO). 2004. Economic and budget issue brief: Fuel economy standards versus gasoline tax. Washington, DC. Online at *http://www.cbo.gov/ftpdocs/ 51xx/doc5159/03-09-CAFEbrief.pdf*.

Energy Information Administration (EIA). 2008. Energy market and economic impacts of S.2191, the Lieberman-Warner Climate Security Act of 2007. Washington, DC: U.S. Department of Energy.

Energy Information Administration (EIA). 2008a. *Annual energy outlook 2008 with projection to 2030.* Washington, DC: U.S. Department of Energy.

Holttinen, H., P. Meibom, C. Ensslin, L. Hofmann, A. Tuohy, J.O. Tande, A. Astanqueiro, E. Gomez, L. Soder, A. Shakoor, J.C. Smith, B. Parsons, and F. van Hulle. 2007. State of the art of design and operation of power systems with large amounts of wind power: Summary of IEA wind collaboration. EWEC 2007 Conference, May 7–10, Milan. Online at *http://www.risoe.dk/rispubl/art/2007_120_paper.pdf*.

Intergovernmental Panel on Climate Change (IPCC). 2007b. *Climate change 2007: Synthesis report.* Contribution of Working Groups I, II, and III to the Fourth Assessment Report of the Intergovernmental Panel on Climate Change, edited by Core Writing Team, R.K. Pachauri, and A. Reisinger. Geneva, 104.

Keohane, N., and P. Goldmark. 2008. *What will it cost to protect ourselves from global warming? The impacts on the U.S. economy of a cap-and-trade policy for greenhouse gas emissions.* New York: Environmental Defense Fund.

Murray, B.C., B. Sohngen, A.J. Sommer, B. Depro, K. Jones, B. McCarl, D. Gillig, B. DeAngelo, and K. Andrasko. 2005. *Greenhouse gas mitigation potential in U.S. forestry and agriculture.* EPA 430-R-05-006. Washington, DC: Environmental Protection Agency, Office of Atmospheric Programs.

Office of Energy Efficiency and Renewable Energy (EERE). 2008. *20% wind energy by 2030: Increasing wind energy's contribution to U.S. electricity supply.* DOE/GO-102008-2578. Washington, DC: U.S. Department of Energy. Online at *http://www.20percentwind.org/20p.aspx?page=Report*.

Pollin, R., H. Garnett-Peltier, J. Heintz, and H. Scharber. 2008. *Green recovery: A program to create good jobs and start building a low-carbon economy.* Political Economy Research Institute, University of Massachusetts–Amherst, and the Center for American Progress.

Searchinger, T., R. Heimlich, R.R. Houghton, F. Dong, A. Elobeid, J. Fabiosa, S. Tokgoz, D. Hayes, and T. Yu. 2008. Use of U.S. croplands for biofuels increases global warming pollution from land-use change. *Science* 319(5867):1238–1240.

Stern, N. 2006. The Stern review: The economics of climate change. London: HM Treasury, Office of Climate Change. Online at *http://www.hm-treasury.gov.uk/sternreview_index. htm*.

Union of Concerned Scientists (UCS). 2009. Clean power, green jobs. Cambridge, MA.

Worrell, E., J.A. Laitner, M. Ruth, and H. Finman. 2003. Productivity benefits of industrial energy efficiency measures. *Energy* 28:1081–1098.

Success Story: Some Good News in Hard Times

American Wind Energy Association (AWEA). 2009c. Wind energy grows by record 8,300 MW in 2008. Press release, January 27. Online at *http://www.awea.org/newsroom/releases/ wind_energy_growth2008_27Jan09.html*, accessed on January 29, 2009.

Apollo Alliance and Green for All. 2008. *Green collar jobs in America's cities: Building pathways out of poverty and careers in the clean energy economy.* Online at *http://www.greenforall.org/ resources/green-collar-jobs-in-america2019s-cities*, accessed on April 6, 2009.

Bureau of Labor Statistics (BLS). 2008. The employment situation. Press release, October. Online at *http://www.bls.gov/ news.release/archives/empsit_11072008.htm*, accessed on January 28, 2009.

Fulton, G., and J. Cary. 2008. *The Michigan economic outlook for 2009–2010.* November 21. Ann Arbor, MI: University of Michigan. Online at *http://rsqe.econ.lsa.umich.edu/michigan_ forecast.pdf*, accessed on January 29, 2009.

Glasscoe, S. 2009. Personal communication. January 29. Sydney Glasscoe is on the staff of Lansing Community College's Environmental, Design and Building Technologies Department.

Goodman, P. 2008. Splash of green for the rust belt. *New York Times*, November 1.

Hemlock Semiconductor Corporation. 2007. Hemlock semiconductor to expand Michigan polysilicon operations. May 2. Online at *http://www.hscpoly.com/content/hsc_comp/ HSC_News_Expand2007.aspx*, accessed on January 29, 2009.

Lee, R. 2008. Personal communication. Rodney Lee is an alumnus of the Richmond BUILD program. For more information see *http://www.ucsusa.org/clean_energy/faces/faces.html*.

New Mexico Business Weekly (NMBW). 2008. GE promises to hire wind technicians trained at Mesalands. May 30. Online at *http://albuquerque.bizjournals.com/albuquerque/ stories/2008/05/26/daily29.html*, accessed on April 6, 2009.

Office of Energy Efficiency and Renewable Energy (EERE). 2008. *20% wind energy by 2030: Increasing wind energy's contribution to U.S. electricity supply.* DOE/GO-102008-2578. Washington, DC: U.S. Department of Energy. Online at *http://www.20percentwind.org/20p.aspx?page=Report*.

Solar Energy Industries Association (SEIA). 2009. Solar and wind ready to lead new clean energy economy. January 9. Online at http://www.seia.org/cs/news_detail?pressrelease. id=322, accessed on January 29, 2009.

Sterzinger, G., and M. Svrcek. 2005. *Component manufacturing: Ohio's future in the renewable energy industry.* Renewable Energy Policy Project, October. Online at *http://www.repp.org/ articles/static/1/binaries/Ohio_Manufacturing_Report_2.pdf*, accessed on January 27, 2009.

Success Story: The Early Feats and Promising Future of Hybrid-Electric Vehicles

Associated Press (AP). 2007. Honda to discontinue hybrid Accords. MSNBC Business, June 5. Online at *http://www. msnbc.msn.com/id/19049079*, accessed on February 4, 2009.

Doggett, S., and J. O'Dell. 2008. Chrysler to discontinue Dodge Durango and Chrysler Aspen hybrids, shutter plant: Move not seen as indicator of a hybrid collapse amid industry's financial turmoil. Edmonds.com, October 23. Online at *http://blogs.edmunds.com/greencaradvisor/2008/10/chrysler- to-discontinue-dodge-durango-and-chrysler-aspen-hybrids- shutter-plant-move-not-seen-as-indicator-of-a-hybrid-collapse- amid-industrys-financial-turmoil.html*, accessed on February 4, 2009.

Hall, L. 2009. Back to the past: The history of hybrids. MSN Autos. Online at *http://editorial.autos.msn.com/article. aspx?cp-documentid=435222*, accessed on February 4, 2009.

Honda. 2009. Insight: Specifications. Online at *http:// automobiles.honda.com/insight-hybrid/specifications.aspx*, accessed on February 4, 2009.

Kiley, D. 2008. Ford fusion smokes Camry hybrid on fuel economy. *BusinessWeek*, December 29. Online at *http://www. businessweek.com/autos/autobeat/archives/2008/12/ford_fusion_ smo.html*, accessed on February 4, 2009.

Ward's Auto Data. No date. Online at *www.wardsauto.com* (by subscription only).

Cultivating a Cooler Climate: Solutions That Tap Our Forests and Farmland

Baker, J.M., T.E. Ochsner, R.T. Venterea, and T.J. Griffis. 2007. Tillage and soil carbon sequestration—what do we really know? *Agriculture, Ecosystems and Environment* 118:1–5.

Blanco-Canqui, H., and R. Lal. 2008. No-tillage and soil-profile carbon sequestration: An on-farm assessment. *Soil Science Society of America Journal* 72(3):693–701.

Climate Change Science Program (CCSP). 2007. *The first state of the carbon cycle report (SOCCR): North American carbon budget and implications for the global carbon cycle.* Asheville, NC: National Oceanographic and Atmospheric Administration, National Climatic Data Center.

Congressional Budget Office (CBO). 2007. *The potential for carbon sequestration in the United States.* Publication no. 2931. Washington, DC.

Forest Service (USFS). 2007. *Interim update of the 2000 renewable resources planning act assessment.* Washington, DC.

Khan, S.A., R.L. Mulvaney, T.R. Ellsworth, and C.W. Boast. 2007. The myth of nitrogen fertilization for soil carbon sequestration. *Journal of Environmental Quality* 36:1821–1832.

Murray, B.C., B. Sohngen, A.J. Sommer, B. Depro, K. Jones, B. McCarl, D. Gillig, B. DeAngelo, and K. Andrasko. 2005. *Greenhouse gas mitigation potential in U.S. forestry and agriculture.* EPA 430-R-05-006. Washington, DC: Environmental Protection Agency, Office of Atmospheric Programs.

Searchinger, T., R. Heimlich, R.R. Houghton, F. Dong, A. Elobeid, J. Fabiosa, S. Tokgoz, D. Hayes, and T. Yu. 2008. Use of U.S. croplands for biofuels increases global warming pollution from land-use change. *Science* 319(5867):1238–1240.

Smith, J., and L. Heath. 2004. Carbon stocks and projections on public and private forestlands in the United States, 1952–2040. *Environmental Management* 33:433–442.

Stein, P., C. Bales, S. Brown, T. Clay, B. Gentry, B. Ginn, J. Horowitz, P. Iwanowicz, M. Jenkins, W. Klockner, D. Liebetreu, A. Lucier, J. Scott, M. Virga, and L. Wayburn. 2008. *Forestry and land-use: Report of the American response to climate change conference.* Tupper Lake, NY: The Wild Center.

van Mantgem, P.J., N.L. Stephenson, J.C. Byrne, L.D. Daniels, J.F. Franklin, P.Z. Fulé, M.E. Harmon, A.J. Larson, J.M. Smith, A.H. Taylor, and T. Veblen. 2009. Widespread increase of tree mortality rates in the western United States. *Science* 323:521–524

Success Story: Farmers and Fungi: Climate Change Heroes at the Rodale Institute

Rodale Institute. 2009. On our farm. Kutztown, PA. Online at *http://www.rodaleinstitute.org/on_our_farm*, accessed on February 2, 2009.

Chapter 8
The Way Forward

Boucher, D. 2008. *Out of the woods: A realistic role for tropical forests in curbing global warming.* Cambridge, MA: Union of Concerned Scientists.

Kindermann, G., M. Obersteiner, B. Sohngen, J. Sathaye, K. Andrasko, E. Rametsteiner, B. Schlamadinger, S. Wunder, and R. Beach. 2008. Global cost estimates of reducing carbon emissions through avoided deforestation. *Proceedings of the National Academy of Sciences* 105:10302–10307. Washington, DC.

Rudel, T.K., O.T. Coomes, E. Moran, F. Achard, A. Angelsen, J. Xu, and E. Lambin. 2005. Forest transitions: Towards a global understanding of land use change. *Global Environmental Change* 15:23–31.

Photo Credits

Covers

iStockphoto.com (blueprint); Jupiter Images (stock exchange); PPM Energy (turbines); Travel Portland (light rail); NYSERDA (insulation installer); Steve Hall/Hedrich Blessing (building); Associated Press (car sticker)

Executive Summary

Page x	Clipper Wind Inc.
Page 1	iStockphoto.com (traffic); Jupiter Images (landfill); PhotoDisc (power plant); PhotoDisc (industry); NREL (kids with bus)
Page 3	Suzlon Wind Energy Corporation
Page 5	iStockphoto.com
Page 6	ACCIONA
Page 7	NREL (turbine); iStockphoto.com (houses); Jupiter Images (office towers); Index Open (gears)
Page 8	Index Open (gears); iStockphoto.com (traffic); iStockphoto.com (houses); Jupiter Images (office towers); Travel Portland (light rail)
Page 9	Sui-Setz (train); David J. Moorhead, University of Georgia (pines)
Page 10	Jupiter Images (stock exchange); iStockphoto.com (U.S. Capitol)
Page 11	Associated Press (man); iStockphoto.com (White House)

Chapter 1

Page 12	iStockphoto.com
Page 14	iStockphoto.com (traffic); Jupiter Images (landfill); Jupiter Images (power plant); PhotoDisc (industry)
Page 15	Doug Boucher
Page 16	NASA
Page 17	Index Open (polar bear); NOAA (hurricane); NASA (Earth)
Page 19	Dennis Marsico

Chapter 2

Page 20	iStockphoto.com
Page 22	NYSERDA
Page 23	Aaron Ogle
Page 24	Picturequest
Page 25	iStockphoto.com (lightbulb); Associated Press (car sticker)

Page 26	NREL
Page 27	NREL (left); Community Energy Inc. (right)
Page 29	UNH Photographic Services
Page 30	iStockphoto.com
Page 32	Dr. Timothy Volk

Chapter 3

Page 34	Chicago Department of Environment
Page 36	Jupiter Images
Page 37	iStockphoto.com
Page 38	NREL
Page 41	Creative Commons (Philadelphia); Jupiter Images (smokestack)

Chapter 4

Page 44	Timothy Hursley
Page 46	Brooks Kraft
Page 47	NREL
Page 48	iStockphoto.com
Page 49	iStockphoto.com
Page 50	PhotoDisc (gas burner); Jupiter Images (light switch)
Page 52	Detroit Shoreway Community Development Organization–Cleveland EcoVillage (left); Southwest Minnesota Housing Partnership (right)
Page 54	NREL
Page 56	Steve Hall/Hedrich Blessing (left); Scott McDonald/Hedrich Blessing (right)
Page 57	Ecopower
Page 59	Rochester Business Journal
Page 61	NREL (all)

Chapter 5

Page 62	NREL
Page 63	PhotoDisc (cooling towers); PhotoDisc (smokestack); Jupiter Images (dam); PhotoDisc (gas burner)
Page 64	PhotoDisc (top); Puget Sound Energy (bottom)
Page 66	Rachel Pachter
Page 67	PPM Energy
Page 68	(top to bottom) Community Energy, Inc.; NREL; NREL; Calpine Corporation; Low Impact Hydropower Institute
Page 69	Make It Right Foundation–Concordia (house); Rachel Pachter (turbine)